AF797880

RANDOM COSMOS

Written by Rajesh` in the company of a little bit of courage, moderate amounts of dedication, and one too many epiphanies in the beautiful city of Bengaluru.

ISBN: 978-93-5361-527-7

First published in 2019.

All content in this book is copyright [© 2019] of the Author, apart from where otherwise stated.

All rights reserved. No part of this book may be used, reproduced, distributed, or transmitted in any form or by any means whatsoever without a written permission from the Author, except in the case of brief quotations embodied in critical articles and reviews, and for other uses permitted by copyright law.

The Author is solely responsible for the book's content, including, but not limited to, the opinions, representations, statements, and references. None of the content of this book shall be construed to reflect the opinion or expression of the Publisher.

Cover page image: pexels.com

FOR YOU

Out of all the infinite possibilities that can take form and exist, you are one. And so am I.

You and I are related in ways we know little about.

NOTE

This book, by no means, is intended to prove anyone right or wrong, or hurt the sentiments of any religion, community, or group. This book is written devoid of the human creations of hate, prejudice, and bias. All views expressed are my own and I am not an authority on any subject. You have free will to agree or disagree with my views.

Please do note that all generic references to the male gender (man, men, etc.) apply to all genders.

While this book has been published with all sincere efforts taken to make the material error-free, I would be happy if you could point out any errors. After all, we are imperfect beings living in an imperfect universe. That being said, 13 grammatical typos have been deliberately and carefully inserted all across the book, for you to find, collate and then send back to me. You stand a chance to win a free copy of my next book (if there is one).

Connect with me on Twitter @sapienRajesh

CONTENTS

INTRODUCTION

I have often wondered where the universe came from, why reality is the way it is, and how did something like life begin. This book is an attempt to present some views on the riddle and purpose of existence. I have strived to work my brain to process all that my senses can perceive and subsequently put forward a perspective and argument on some interesting topics that I have come across. Also, there are various theories, principles, concepts, philosophical arguments, laws, and observational data that I have researched and presented in brief, to put together a comprehensive narrative for you. None of them are overly detailed, as this is not intended to be a scientific journal.

Do not mistake this for a book on the story of the cosmos. I have tried to make this into a casually structured conversation-style narrative on subjects you'll find worth reading, and you'll discover questions scattered everywhere in the book. These questions are intended to make you pause for a moment, process answers your mind can come up with, and then proceed. Deep dives into any of the theories or principles have been avoided to not make us digress from the overall theme, which is to look at the what, how and why of all that's around us. What that also means is that unless you are familiar with science as a subject, you might need to, on some occasions, do some searching on the Internet if you want to know more about some scientific concepts that have been referred to. This is optional and should you choose to skip it, it won't take away any of your takeaways from this reading effort. There are imaginary conversations thrown in at some places, with a lot of thought, to add intellectual substance and provoke your thoughts. I have also briefly touched upon the lives of many famous and infamous humans, that I am sure you will find interesting. There are no photos, images, graphs, or charts of any nature. It's all straightforward text and your mind is free to paint pictures as and when it pleases.

I believe that growing up is a process that happens throughout our lives. Every new day is an opportunity to try, experience and learn something. It is also never a wrong time to unlearn something that no longer makes sense. It has been a personal quest of mine to get on the road to truth and aim for wisdom. In the course of that pursuit, I have found that the road to truth is choked with lies masquerading as truth. I have sought to unmask and present such false concepts in this book, for your neutral consideration. My quest to find the answers to the most profound questions has also led me to discover and understand the works of many others who have researched great mysteries of the cosmos. I am truly inspired by the works of many such distinguished thinkers, to whom I am grateful for making me wiser and guiding me to reality. I am amazed by the astonishing depth of thought of fellow human beings who have walked this land. Honestly, this is such a grand and fascinating subject to study that I often wonder how one can return to the trivial nature of everyday existence and chores after contemplating such profound topics.

You would agree with me that it takes courage to pen down a book, especially one that is non-fiction and opens the author's mind to be viewed and examined by all readers. Needless to say, including topics like religion is always controversial. I find the courage and take this risk because I feel the need to. I studied science growing up, then took up a career in corporate human resources for a long time, and am now exploring a professional career in finding the absolute cosmic truth. It is funny that only when I stopped doing anything scientific did I really start understanding science. It was not my life's purpose to write a book, nor is it a checklist item on my bucket list. At this juncture in my life, I find my calling in writing something of this nature. Sharing with you about what I have understood about everything is what makes sense to me right now. I am both a debutant and a dilettante. Where observed, do pardon my ignorance. I may be guilty of ridiculing something I think I understand but perhaps I don't. But where deemed fit, do appreciate my argument, insights, and point of view. The only sensible thing to do is to not be stubborn. Let's get along and find truth & meaning together.

Nothing that we see around us is by design. Everything is random, by chance, through natural evolution via accidental mutations and the rule of survival of the fittest. The result is that we are here, to figure it all out. One of the things most have figured out is that we are not a victim of karma, destiny, design or anything of that nature. We make conscious decisions every moment. Always make a conscious decision to not judge a book by its cover, or by reading it. Just read it with an open mind, seek knowledge or inspiration if there is any on offer, make room for realization and appreciate the efforts of the writer. Billions of people have lived on this land and the language dictionary is sufficiently limited. What I think and write may have been thought and written before, but I hope that this book, in its overall presentation, will put forward a fresh and engaging perspective on the riddle of existence and what we are supposed to do with the gift of life.

I look forward to your patience, time, and attention as you voyage into this book. Happy reading!

UNIVERSE & CREATION

The most beautiful thing we can experience is the mysterious. It is the source of all true art and all science. He to whom this emotion is a stranger, who can no longer pause to wonder and stand rapt in awe, is as good as dead: his eyes are closed.

– Albert Einstein
Living Philosophies (1931)

One cannot not be overwhelmed by the magnificence of all creation. Scientific research has found a visualization medium in the form of as-good-as-real computer graphics to show us, with great clarity, how the universe clearly works. It has made us see how there was a supposed big bang at the beginning, how matter got together, and how galaxies, stars, and planets were formed. We have learnt that there are basic elements that exist in the universe and everything is made of those. You and I know that we live on planet Earth, in our solar system centered on the Sun. Our solar system sits in the suburbs of the Milky Way galaxy, which is one among an estimated 100 billion galaxies in the observable universe. While we try to come up with the number of zeros there are in a billion, we ought to wonder as to where did all of this come from? Who designed and created all of this, and then put us here to figure it all out?

Around 2,500 years ago, the great Greek philosopher Plato came up with his theory of reality called Platonic realism. He pointed out that all visible things in this world are like shadows cast on a wall by the activities of their corresponding universal forms. He said that our visible world of things you see is not real, and it is actually those universal forms that occupy the unobservable yet true real reality. This is truly quite a complex thing to comprehend and believe. However, as you will observe eventually, in all the complexity also lies the beauty of everything. Some philosophers think that the universe is just a thought experiment and once it gives rise to conscious beings, they bring this experiment to life. There are loads of such philosophies and other philosophical ideas attempting to explain the origins of the universe. These present interesting abstract theories that seem to answer all questions one can have, but most theories do not tally with any scientific evidence at hand.

In Hindu philosophy, Brahman is considered to be the cause of all that exists and the highest universal principle and reality. It is the formless spirit from which everything comes. It is talked about as the infinite, eternal and unchangeable truth. Old sages called it as the sole reason why things in the universe exist and change. Apart from testing the authenticity of it, the other concern with any old philosophy is that it is old. We all know how we have been proven wrong in the past. For over a thousand years, all the intellectual minds shared the commonly agreed and no-brainer understanding that the universe revolved around an unmoving Earth.

The cosmos can humble you like no other. What other no-brainer concepts that exist today will be proved nonsensical in the future? History has proven that, contrary to collective convincing understanding at certain periods in time that all has been figured out, the past is mostly recollected as an age of ignorance and delusions. Future civilizations will think of us and our beliefs the same way. At the same time, I doubt if they will doubt our intellect when it comes to some of the questions we have asked and are still searching answers for. The deepest and most intense cosmic question is why there is 'something' rather than 'nothing'. Wouldn't it be a simpler and more natural state if nothing, no space or time, existed? This question was put forward by the German philosopher Gottfried Leibniz around 300 years ago and is referred to as the most fundamental question of metaphysics. The question is both illogical and deeply profound at the same time. Think about it, why should anything go through the bother of existing? On this subject, it is pertinent to think about the fact that how 'nothing' could have resulted in 'something' that started the universe. Unless, of course, it turns out that everything is eternal, which itself is a hard to fathom concept for the logical mind. Nevertheless, can something arise from nothing? To understand how it cannot, we must understand causality.

Causality is the unchangeable relationship between a cause and an effect. An effect is because of a cause. You are reading this sentence because it's printed in this book. You cannot read an empty book. Nor you wanting to read will create content in a book. An effect cannot create a cause. Always remember this most basic of concepts. So, what caused the first cause in this cycle of causes and effects? Well, nobody knows. Unless, of course, 'nothing' was the first cause because of some phenomenon we are yet to understand. But then, in the presence of that phenomenon, how can we say that there was nothing? And, what caused that phenomenon in the first place? We can go on and on, turning it into an infinite regress. We should also know that science cannot accommodate the concept of perpetualness or eternity because it violates the fundamental cause and effect law and hence pretty much breaks down most of our understanding. There is nothing eternal when it comes to science.

We'll need to refer to a whole lot more of science as we go along. It makes sense to introduce cosmology at this stage, which is that branch of astronomy that studies the origin of the universe, and how it has evolved from its birth and into the future. It does so using scientific observations, experiments, inferences and a few assumptions thrown in for good measure. In my opinion, cosmology is the quest to find the ultimate truth. What is truth, you ask? Truth is that which is in accordance with fact or reality. Okay, what is reality? Well, that I don't know for a fact. What I do know is that it is the pursuit of truth that drives us. Will we be able to find this truth in this book? Hold on, we seem to be getting ahead of ourselves at this point. We'll find out answers to those questions in due course of time, and space. For now, here are some random facts from a science book: our galaxy is constantly moving, as the universe expands. The Sun and our solar system travel with our galaxy. Our solar system travels at an average speed of 800,000+ km/hr. We should know that nothing in the cosmos sits still although everything appears to be stationary. In stillness lies all motion. In poetic terms, we're all passengers on a journey who are forever traveling, but no one knows to where. Perhaps I'll be able to shed some light on probable destinations, a little later in this book. Stick with me. Won't you!

We want to read the story of creation and we want it to be the right one. The universally accepted theory for a while now has been the Big Bang theory. Most of the aspects of the theory have been scientifically proven over time, most recently with the cosmic microwave background readings that clearly show the leftover radiation from that early explosion. So, our observable universe did indeed have an explosion at some early point and it then increased in size with time. However, what we will never be able to figure out is how it all started. What caused this explosion or bang? I am not ready to accept the hypothesis that empty space and time self-erupted into existence, out of sheer nothingness at the moment of singularity or creation. I simply cannot wrap my mammalian brain around the notion that there can be a reality where there is no space and time. Can you?

Accepting the Big Bang theory officially and yet surrendering that we cannot know what caused the initial bang is equivalent to being born but not knowing how you were born, were there parents involved, who were they, and how and where were they born? That just won't do. We cannot accept a theory that does not answer the most basic of questions. The notion that the big bang happened without a cause is as far-fetched as a divine creator creating the universe in a single breath. It may well be that our universe is cyclical and it underwent a collapse (Big Crunch theory) and then restarted through an explosion. We may never know. We'll obviously never know what started this cycle in the first place. In the great library of the cosmos, important volumes on its birth and death are blank. This reminds me of a conversation I dreamed of a while back:

Man: I feel a void within me sometimes. It's like the nothingness that exists in the universe has seeped into me.
Wise teacher: I can relate to your feelings of emptiness. Always remember that the universe works on a set of basic principles. The principles are not hard to understand, but the fact that everything was created on such basic principles baffles many, including me.

Man: How can the universe be eternal and have no beginning? Nothing else that we have seen in the universe is eternal. Everything has had a beginning and an end.
Wise teacher: The forces of gravity, electromagnetism, etc., are also eternal in that sense, aren't they?
Man: Well, okay those also. But why would the universe, which itself is eternal, create time-bound phenomena?
Wise teacher: I don't think the universe has consciousness to create phenomena intentionally. Also, isn't time eternal if the universe is eternal?
Man: It is time to go for a drink.

One of the great mysteries of how human beings came into being was resolved by the English naturalist Charles Darwin in the 1850s through his painstaking discovery that inherited traits that enable organisms to survive and reproduce become more common in a species over time. In simpler words, only the fittest survive and grow while others die out with time. This is called natural selection. Organisms undergo random genetic mutations as they reproduce, and thus develop traits that help them adapt to their environment better than others. As we will learn in the chapter on evolution, all organisms have descended from a common ancestor. This all started a few billion years ago when that first lonely cell chose to divide into two. The rest is history, as they say.

I know what you are thinking. That this chapter so far is largely an unstructured collection of random facts and thoughts. Well, kind of yes. You see, I am warming you up for what is to follow. The flow of the cosmic energies may not be by design, but the flow of this chapter surely is. We move ahead. I am sure that the first thing a curious mind wants to figure out is how it came into being. But, however hard I may try, I can't remember the most important day of my existence: that day on which I was born. On that subject, I realize that I won't have a working brain to recollect the day of my death. So, my brain is useless when it comes to the two most spectacular days of my life. It's as if your camera stopped working when the white tiger appeared out of the woods. As that may be, the fact remains that we were all born on some day in some year. But that's our biological birth date. While I had no say in my being born, I am interested in figuring out when was I truly born? Did I come into being on:

- the day I came out of my mother's body?
- the day I was conceived by my parents?
- the day the seed of my birth (sperm/ovum) was generated in the body of my parents?
- the day my parents were born?
- the day my parents were conceived?

For my parents to be conceived, my grandparents had to be born. If you follow the same logic back to the first human, I owe my birth to the birth of the first human, who in turn came into being after billions of years of evolution, which happened because of the specific set of properties that planet Earth possessed. Planet Earth itself was formed when the solar system was conceived, which was the result of some star that died billions of years ago. The star came into being because the universe created stars through nebulae nurseries, as a result of its own basic construct and properties. So, I thereby conclude that I came into being the day the universe came into being. I owe my existence to it. When I die, I will dissolve into the world that I lived in. The basic elements and energy that made me will go back to the universe that made them, to be used to create something else. If the universe started in a big bang, then that's where I started. If the universe is indeed eternal, then my essence and being did not have a beginning and will have no end. Only the epoch during which I took the form of a human with consciousness is time bound. Nothing else about my being is. I may be eternal, and so could you. But we don't seem to have sufficient time to find that out.

I love to gaze at the night sky and imagine how everything must have happened. I like to think that a spaceship waits for us every night. It does not take us to space but brings space to us instead. We just need to gaze at the sky and wonder. We can also choose to think of the Earth as a vehicle. This vehicle rotates 360 degrees and revolves perpetually around a center, both at phenomenal speeds. The sky above is our window, offering magnificent views during the journey. No one got motion sickness from watching those. While we enjoy the sights, we need to keep in mind that we are looking into the past when we look at the objects in the night sky. That's because the speed of the light that hits our eyes is finite (300,000 km/sec) and our galaxy, the Milky Way, is huge (9.5×10^{17} km in diameter). Thus, the light that's carrying images of these objects takes a lot of time to reach us. The night sky is, in a way, a collection of recorded footage. Some recorded a few minutes back and some millions of years back. None of it is in the present. Some of the stars we see may have already died a long time ago. Betelgeuse in the Orion constellation is one such star. It shines every night but probably is no longer alive. There are no ghosts on the land, but the sky is full of them.

The universe is full of carbon. It is everywhere. It also has humongous amounts of other basic elements like hydrogen, nitrogen, oxygen, and the omnipresent natural forces like gravity and electromagnetic force. All of them have worked together to create all the matter in the cosmos. Elements like hydrogen and helium were formed in the Big Bang explosion and the remaining elements came to life in nuclear reactions inside stars and when they died through giant explosions. The natural or physical forces that enabled all of this ought to have started when the universe started, as they were not created because of some phenomenon. Rather, they enabled the phenomenon of matter clumping together and forming the stars and elements, which in turn formed everything else. So, why are the natural forces the way they are? Why does gravity exert that precise amount of force that it does? Should we treat them like divine commands or plain brute facts that have no explanation? The mystery of when, why and how natural/physical forces and laws came into being is what still connects science with theology (i.e. God and religious beliefs). Unfortunately, neither of them has provided any convincing arguments to support their claims. Science tells us that the physical laws that apply as brute facts in our universe are not divine commands, as our universe may be one of countless many (what!). It further tells us that each of these universes has a different set of physical laws, which make many of them habitable and many absolutely not. But what starts this process of creation of these multiple universes is still unanswered. Theology, on the other hand, has nothing to do with any of this nonsense and attributes everything easily to God's design and will.

Easy to find things are not worth looking for. That being said, we're not going to endlessly ponder over our origins via this book, as there are countless initiatives and efforts on for generations that are focused on that. Although, religion continues to maintain that it is God's design and we should accept it as it is. The Vedic text Kena Upanishad emphatically says that the Brahman cannot be comprehended, because it has no attributes and is an unthinkable, indescribable, eternal and an all-present reality. In any case, let us dip our toes a tad so that we have sufficient knowledge to understand what is to follow. We want to understand how everything works. Fair enough. At the large scale, we have general and special relativity theories to help us understand that. Among other revelations, they talk about how gravity arises from the curvature of space and time. Curved four-dimensional space-time is quite impossible to imagine. But, as it turns out, that is the space we all live in. At the microscopic level, quantum mechanics helps us understand the behavior of particles like electrons. Interestingly, the word 'atom' is derived from the Greek word 'atomos', which means indivisible. The atom, however, turned out to be further divisible into protons, electrons, and other smaller particles. The name atom, however, was retained. In any case, as it turns out, both of these sciences are not consistent with each other. Important things like gravity are explained differently. This thus leads us to believe that the real correct theory must be something else, for it should apply to everything in the universe consistently. So, have we been wasting time in trying to unify two incorrect theories? Will we ever get to know? One fascinating thing we have got to know about quantum physics is that the presence of an observer changes the observation. It's like saying that quantum particles know once someone is observing them and electrons get conscious in our presence. Imagine the contents of this book changing on their own depending on who is reading the book. Consciousness collapses the quantum wave function, as scientists like to put it. Wow, let's park that for a while and leave those wave functions alone. No peeping allowed.

Man: I want to give up. My sensory organs and my brain have evolved just enough to ensure that I survive and reproduce. They are not equipped to experience higher reality and answer profound questions, as that is not a survival need.
Wise scientist: That is actually true. But tell me what are you looking for?

Man: I am looking for something that applies to everything. In simpler words, I am looking for the truth.
Wise scientist: Where are you looking for it?
Man: At present, I am looking for it in quantum theory.
Wise scientist: All of reality is probably nothing but a quantum wave function. Our universe may be the play of quantum fluctuations. But where did those laws come from, which allow those quantum fluctuations? The limitation of logical scientific explanation is that it has limits and cannot answer everything. Physicists cannot answer where the forces like gravity came from, or how time started.
Man: So, what do you suggest I do?
Wise scientist: Well, let us keep looking everywhere. We won't find it in one place. We may probably find one clue or message here and there. It is for us to accumulate the various parts and bring them together to seek the truth.

There is not much common jargon that both the scientific and the mystical minds employ. Matter and energy are some of the few that are contemplated about by both science and religion. They were perhaps created once upon a time or are forever eternal. In either case, they are here. At a conceptual level, we can view energy as the abstract representation or effect of the natural forces. It's a little difficult to explain without getting into nerdy equations, but thankfully you are smart and already get it. The proven laws of conservation tell us that energy is going to be here always and in the same net quantities. If energy cannot be created or destroyed, then the sum total energy in the universe must be zero. The net energy cannot be positive or negative. That would violate a lot of our understandings.

On some mystical mornings, I think that matter and energy are immortal and will be there forever. They take up temporary forms from time to time. You and I are one such temporary form. On some philosophical evenings, I cannot help but think about how abstract energy is. Sure we can measure it. But how do we define it? We have negative forms of it in the form of potential energy. We have positive forms of it in the form of kinetic energy. Light has it, heat has it, sound has it, and all matter has it. Also, it cannot be created. In some sense, it is like a wandering spirit that keeps changing its form from time to time. Some of us can sense negative energy in other humans. What is all this? On some nights when I don't feel like stepping out yet lie restless at home, I've wondered on the astonishing marvel that physical laws, matter, and energy—all of which are devoid of any consciousness, have come together to create supremely aware beings such as us. Is intelligent life the most astonishing product of the works of them? Or, has the awesome trio built something even more incredible for us to look forward to finding? Perhaps they built intelligent life just so that it could witness the next level of creation. But the trio is not aware and intelligent on their own to have done this intentionally. Let us divert our focus and attention from this deep and perplexing subject for the time being. A bad headache can destroy all plans for the day.

The secrets of the universe are written in a language we cannot understand, printed in a book we cannot find, however hard we try. Do our thoughts have a language? What is language anyway and how did it emerge? This has been a subject of speculation for many centuries and has been vehemently debated. So much so that sometime in the 1860s, the Linguistic Society of Paris banned any existing or future debates on the subject! Language is a method of communication, and that, in turn, is the process of sharing of our brain impulses via our sensory organs. There are currently around 7,000 languages and dialects in the world. We communicate through language, both verbal and written. We also communicate without language, all of which is shared, received and interpreted by our five senses. The sense of sight is the strongest among the five. Sometimes I wonder if language affects reality? The linguistic relativity (or Sapir-Whorf) hypothesis indicates that the structure of a language affects its speakers' cognition. So, in a dark world where the word happiness does not exist in the language or the dictionary, does the emotion then lose its existence? I don't know. Perhaps some answers can be heard in silence. Also sometimes, no words are required to get the message across. The Persian poet and mystic Rumi once said, "Silence is the language of God; all else is poor translation."

All languages, but one, are man-made. The exception is mathematics. Mathematics is a naturally occurring universal language. It is true, whether you believe in it or not. It will exist, independent of our existence. All future space travelers should have a command on it, so as to communicate easily with other intelligent life forms, should that need arise. That being said, if math is indeed the language of the universe, then it should be hard-coded into all living species just like how breathing is. Also, evolution would have eradicated the beings bad at math. So, that makes all living species in this age very good at math skills required for survival and reproduction. Including birds, bees, and whales.

Language has allowed great minds to communicate freely with each other, and it is from them that we learn that the most abundant things in the cosmos are emptiness, darkness, and silence. Ironically, none of us want those things in our lives. That's okay. We are not perfect beings. In fact, nothing is. As science has observed and understood, in a perfect universe there would be equal amounts of matter and antimatter, which would then balance out leaving nothing but empty space and

DNA would copy itself perfectly 100% without any mutations. Thus leading to no matter, no evolution and hence no you, no me and nothing of all that we see around us. On this subject, it is also pertinent to understand that the early universe just happened to have the sufficient amount of deuterium in order to delay the fusion reactions in the cores of stars, thereby enabling the formation of rocky planets like ours. We owe our existence to imperfection. Let's not try too hard to be perfect in a far from perfect universe.

In a way, our brains mimic the universe in the way they have turned out. They are far from being perfect, not fully evil yet not fully good, not fully brilliant yet not fully naive. But, is perfection an abstract concept made by man and our understanding is simply not perfect enough to realize the actual perfection of the universe? I know, I don't seem to be going anywhere with this. So what, I ain't perfect. Also, all perfection is flawed. Enough with this already. Nobody likes dealing with wannabe perfectionists. In musical terms, there is only so much a classically trained voice can be listened to.

Do dancing bears in the circus muse on the existence of the universe? Let's broaden the scope to include all the near 9 million species of unique organisms discovered so far on this planet. As a matter of fact, a vast majority of the total number of species has not yet been identified, and classifying them all could take a very long time. Some believe that the actual number of species is more than a trillion! In any case, whatever that number may be, what concept of existence and universe do the other species, apart from us, carry? How do they envisage the world started? Don't tell me that they are not intelligent enough to ponder over such subjects. All living species on our planet are intelligent, for they have managed to survive, communicate, learn, build and grow amidst the changing backdrop of nature. The degree of intelligence clearly varies.

Humans are the most intelligent among all, through the natural evolution of our brains and improvements in them through our accomplishments over eons. But others species also possess a sufficient amount of intelligence. I am inclined to add that they are in some ways smarter than us. If you've noticed, animals have not bettered their shelters over time. They still live the same way as their ancestors did. They made simple and effective nesting habitats and stuck with simplicity. On the other hand, we humans seem to have complicated our lives by living in homes and comforts we cannot afford and do not need. But that's just one in our own possible trillion shortcomings.

We evolved just like how other animals did. It's just that our brains got bigger and more evolved than others. So, our focus expanded from just meeting survival needs. Life became complicated. We'll deliberate this at length in the next chapter. For now, we should ponder over what other organisms think about creation and existence? If they are looking for answers, we should share our findings with them. Can we communicate with ants and enlighten them about space-time, gravity, and the origin of life? No, not yet. But, as a fellow species sharing a home with us, they deserve to know. Probably, one day we will be able to.

Man: Sir, I just wrote a poem. Would you like to hear it?
Wise poet: Always.
Man: In this pathless forest,
I walk but cannot see,
I have no understanding of a destination,
Or what reality is supposed to be.
Wise poet: Excellent. Remember that it is your mind that turns words into a poem. I believe you have now evolved your mind to see reality for what it is.
Man: Thanks. But tell me, what is reality supposed to be?

Wise poet: When you reflect on the miracle of existence, you start appreciating the monumental beauty of the universe. And if the universe is indeed the absolute being, then we are its eyes through which it sees its own magnificent beauty. I know nothing less or more than this.

What is reality, anyway? Reality is the sum total of all that is real and our universe is a part of it. Why did reality take the form that it did? The philosophy of metaphysical solipsism says that yourself or your mind is the only reality and all things around you in this world are mere representations of that, having no independent existence of their own. Now, that's some deep computer simulation type hypothesis to digest. Using a saner lens, from all that we have observed and understood about our reality, we know that there are only two states that exist: existence and nonexistence. We have seen that reality is not limited to what you can see, even when aided with a telescope, as the range of spectrum goes beyond the colors of red and violet. What does the true picture of reality, with all the infrared rays, x-rays, microwave, gamma rays look like to the naked eye of an organism that can see it all?

There is an observable universe that the eye can see, and then there is that part that we don't know anything about. Likewise, there is a comprehensible and empirical reality. Then there is that part we can't even imagine, let alone try to understand. A frequent wave that crosses my thought process is that if a thing like reality can come into existence, anything can. Thanks to the engrossing science of biology, we know that bees see and perceive reality differently. They can see ultraviolet light. But we humans cannot. Similarly, other living organisms differ from us in terms of their sensory organs. Their perception of reality differs from ours. All of this leads to the understanding that reality is subjective. If there are other realities, one can only wonder about what other states could exist. Do we even know what to wonder about? The limited dimensions of the reality we live in limit our imagination. I am so high on the wonder of the cosmos right now. It is a drug that's thankfully legal.

One of the contenders of the unified theory that we were alluding to earlier happens to be String theory. It attempts to model the four known fundamental forces (gravitation, electromagnetism, strong nuclear force, and weak nuclear force) together in one theory. String theory tells us that reality consists of tiny vibrating strings in a higher dimension. It also tells us that there are 10^{500} solutions to it, therefore there must be 10^{500} universes or realities, out of which ours is just one. There are other realities where the fundamental forces play out differently. Thereby leading to some universes that are just empty, some with life and some without, and various permutations and combinations of things we are yet to know. But String theory has been around for many decades, yet has failed to prove anything convincingly. So, the jury is still out on that hypothesis. The jury, however, has passed judgment on the fact that reality has dazzled us and shown us everything in great splendor. We don't have anything that can match reality. But, reality also hides a dark secret that it refuses to share, which happens to be the secret of the reason for its own existence. It may be that its marvelous show is an act of overcompensation to hide that dark secret! Look at us. Talking of reality as if it were a conscious being with flaws.

The search for alien life is possibly an area of interest for the world that is on par with the meaning of reality. For long, we have looked for evidence of extraterrestrial life. With time, the very concept has taken the form of a cult with its set of believers and non-believers. From seeing clues in cave art inked thousands of years ago, to actual testimonies of people who claim to have been abducted by tiny green men, to large-scale conspiracies accusing the government of destroying evidence, this field has captured the imagination of people from all walks of life. It captivated me when I was very young. Indeed, it is the most basic question of curiosity on whether the universe that made us also made others. One such captivated human, the American astrophysicist Frank Drake, wrote this equation back in 1961:

$N = R_* \times f_p \times n_e \times f_l \times f_i \times f_c \times L$

where:

N = the number of civilizations in our galaxy with which communication might be possible (i.e. intelligent alien life), and

R_* = the average rate of star formation in our galaxy

f_p = the fraction of those stars that have planets

n_e = the average number of planets that can potentially support life per star that has planets

f_l = the fraction of planets that could support life that actually develop life at some point

f_i = the fraction of planets with life that actually go on to develop intelligent life (civilizations)

f_c = the fraction of civilizations that develop a technology that releases detectable signs of their existence into space

L = the length of time for which such civilizations release detectable signals into space

(equation explanation is courtesy of Wikipedia)

The last four parameters are not known and quite impossible to estimate with precision. This equation has no correct solution that will tell us the number of alien civilizations in our galaxy. Its purpose lies in helping us deliberate (perhaps philosophically) the factors involved in predicting the presence of such alien civilizations, the time period of their existence, and our ability to make contact with them. Nevertheless, it would be foolish to say that we are the only ones here. Surely, there are other places in the universe that are as fortunate as us to have intelligent and aware beings that can ponder their origins and appreciate the grand design. As the American astronomer Carl Sagan once famously said, "The universe is a pretty big place. If it's just us, seems like an awful waste of space."

I think that given the origin of life on Earth from basic elements that are present everywhere in the universe, life has to exist in many other places. That life may be basic, as advanced as us, or far more advanced than us. It all depends on the kind of environmental conditions and evolution complexity. Sadly enough, we will never know given the immense size of the universe. These islands of life are so vast apart that they will never be able to get in touch with one another. Perhaps the only ones that are within reach and we can meet would be primitive bacteria type life residing in places like the waters of Saturn's moon Europa (that is itself a shot in the dark). No actual intelligent aliens are going to come visit us nor are we going to pay them a visit. Our search for alien life rests on not finding them, but their homes. The quest to find rocky Earth-type planets in other planetary systems is our best bet to speculate about alien civilizations. But not meet.

How do we measure up against other probable intelligent civilizations in the universe? How intelligent is that civilization living on a planet far away? Our depictions of alien species always somewhat resemble us humans in form and shape. Why is that? We are the way we are because we have evolved on this planet and are well suited to this environment. An alien species would take the form and shape via evolution that makes it well suited to its own environment. We can only wonder and imagine the possibilities that exist. Also, we know that there have lived many hundreds of millions of species of living organisms on this planet (including both current and extinct). Out of all of those, only one species was able to figure out a way of sending signals and vehicles to outer space. Isn't it a tad funny then that our general perception of alien life is of a species equally or more developed than us? Our knowledge of probability should tell us that it is way much more possible that other worlds could be inhabited by creatures simply incapable of receiving and decoding our signals, let alone coming up with a spaceship to come visit us. In any case, it would do everyone a world of good if, by some miracle, we could meet an intelligent extraterrestrial life form. For starters, that would help shift our focus from everyday gossip, religious and political wars to something grander. Also, an advanced alien species would be able to help validate our findings and inferences about the workings of the cosmos. A second opinion always helps. We know that the bats and penguins in our world have refused to share their revelations. Unless, of course, aliens turn out to be sinister beings. In one wild theory, it is believed that aliens created humans and then science to ensure that all scientific proof leads to the belief that aliens cannot reach us!

Once we are done with grappling with our loneliness and the ensuing depression of not being able to meet an alien civilization in our lifetimes, we should focus on understanding whatever we can in our limited lifetimes. We may be alone but will always have the company of a countless number of things and concepts to explore and understand. The infinite encyclopedias of the universe forever await inquisitive readers. The exploration of the universe happens to be the ultimate detective expedition. Each observation gets us to a new place with the next set of clues to be found. The clues are limited in number and the space-time scale to cover is near infinite. That fact is also a little depressing. But then, there is some romance we can look forward to. Since time immemorial, we have been known to be in mad love with the cosmos, and many have been swept off their feet by this romance. The overwhelming beauty of the cosmos can change a man permanently and can humble the most hardened of us men, making us stare in sheer awe.

One such awestruck man was the great German astronomer Johannes Kepler, who discovered the motion of planets and gave us laws that formed the foundation of our understanding of the solar system. As a six-year-old boy, he witnessed the Great Comet of 1577, which was visible across Europe. "Seen through the clouds like the moon," said the witnesses. I wonder on how that spectacle in the night sky must have forever sparked the curiosity of that young boy. Although that comet has never returned to be witnessed again, it inspired him along with many others to understand what was going on in those heavens (as it was understood then). I am sure that they must have later realized about how chance and randomness are woven in the way the universe operates. An example being the accidental nature of reality itself.

Things are not here by intelligent design. They are a result of natural laws playing out on the basic elements in the unknowable fabric of space-time. Does that mean that we are momentary and inconsequential beings in an endless, timeless, and indifferent cosmos? Perhaps. But, let's take heart in the fact that all things that happened in the universe before we were born had something to do with us entering the world. From the creation of space and time, to matter and energy being born, to galaxies taking shape, to the birth of stars and planets, to the creation of single-celled organisms, to the natural evolution of human beings, our forefathers and our parents conceiving us. They all played a part. We are a part of the grand design, aren't we? Honestly, this subject is so profound and overwhelming that I cannot work out what to make sense of. We must honour the universe that is, was and will be. Let's name our daughters Gravity and our sons Carbon.

Here is an excerpt from another dark dream I had one fine dark night:

Man: Why is it so dark here? I can't see anything.
Wise doorman: This darkness is infinitely spread and eternal.
Man: Sure it is. For now, can you help me with torchlight? I can't see anything.
Wise doorman: Well, you obviously cannot see nothing. As there is nothing to see. In fact, the very notion that you can glance at it means it is not nothing. You cannot even think of the concept of nothing. This is because an observer, via sight or thought, brings something into place. So, never say that you see nothing. You are being dead wrong.

What is nothing? How do you define the state of nothingness? We had discussed the German wizard Gottfried Leibniz, who also conceived the hard-to-understand subject of calculus, and was one of the first people to formally pose this question of why there is something rather than there being nothing at all. He was of the opinion that a reality where nothing existed was simpler and easier. Leibniz has been dead for more than 300 years now, but we are still puzzling over how to define the state of nothing. For starters, validating it would require the presence of an observer, thereby nullifying the very concept of there being nothing. In other words, the funny thing about reaching the much debated absolute state of 'nothing' is that there would not be any observer to prove that right. Is vacuum a state of nothingness? Apparently, it is not. A vacuum has things at the quantum scale. Also, there is this thing called quantum fluctuations that can make particles appear out of the vacuum of space. Like something emerging out of nothing.

Proponents of the Big Bang theory employ this phenomenon to explain the first cause. Who knows who's right? Maybe nothing knows.

Many religions preach that God created the universe ex nihilo i.e. 'out of nothing'. I have a lot of problems with that approach. For starters, in the absence of space and time, where and when did its creators carry out the grand design meeting of the universe? The riddle of God's existence will be resurrected in a later chapter, for 'nothing' is seeking our undivided attention for now. If we give it some thought, it should occur to us that indeed, a reality where nothing exists is more likely to take shape if provided with an infinite number of forms to pick from. Can non-philosophers and non-physicists like me experience nothingness? If I close my eyes, my mind is still thinking. So, can we? It seems that we can. A dreamless phase of sleep is our mind's experience of nothingness every day. Our very own moments of non-existence in a way. I guess a state of dreamless sleep is what the universe must be experiencing all the time. It exists, but it does not know.

I sometimes worry if we are doing it all wrong. By studying what we can observe and using the readings to arrive at conclusions. The atoms in my body do not have consciousness. My lungs do not have a heart. My heart does not have a pair of legs. If my constituents are so vastly different from me, then how can we perceive to understand the universe as a whole by studying stars and galaxies? We seem to have understood how things work on Earth and in our celestial surroundings, and deduced laws that we believe apply everywhere in the universe. But then we have also come across things like black holes where fundamental physical laws don't apply, and where foundational concepts like time and space have no meaning. So, what else is out there that's inconsistent with our understanding and conclusions? What if every galaxy has its own set of laws, like mini-universes within the universe? The deeper we try to dig into space, time, and the unknown, the closer we get to where we originally started.

It also makes us wonder what all things are possible. Do we even know what we don't know? If I can survive in the openly-respected yet secretly-hated field of corporate Human Resources for around 14 years without having any relevant education or many of the required skills and 'competencies' (occupational jargon) as well as being forever socially awkward, then anything ought to be possible in this accommodating cosmos. Absolutely anything.

Now then, where were we? "Let there be light"—that's a phrase from the Bible Book of Genesis. What is light exactly? In layman terms, it is what makes things visible. Someone once blasphemously remarked that in the beginning there was nothing, and then God said, "Let there be light" and there was still nothing but now you could see it. I am not going to bore you with the scientific definition of light that includes terms like electromagnetism and photons. I am more interested in bringing to light the fact that light is the ultimate evidence gatherer for all us space detectives. With the help of the properties of its spectrum, we can figure out what distant celestial objects are made of. Most of our understanding of the planets in our solar system is on account of this. Light also has a philosophical aspect to it. Without experiencing sorrow, we cannot understand happiness. On similar lines, without seeing and understanding darkness, we would never understand light. In a world without darkness, the concept of light will have no meaning.

Closely associated with light is the notion of travel, which includes distance traversed through space and also the most mind-boggling concept of time travel. Time travel is an area of fascination for many of us. Some of us are obsessed with looking for travelers visiting us from the future and wasting time on examining the paradoxes therein. While traveling into the past is not possible as per our current scientific understanding, it is, however, possible to travel into the future if we can figure out how to travel close to the speed of light. Come to think of it, the photons in light beams are forever traveling at the speed of light. Do the concepts of past, present and future apply to them? Nothing can travel faster than the speed of light, as per the physical laws of the universe we live in. Even if we were able to somehow figure out a way to travel at the speed of light, it would take us no less than 2 million years (!) to reach our nearest galaxy, Andromeda. And that is just one among the 100 billion galaxies in the observable universe.

For us ambitious galactic travelers, exploration of other galaxies in person will forever be a dream. Hey, hold on for a second. I was just able to think about a galaxy that is around 2 million light-years away. (A light-year is the distance traveled by light in one year.) Surely, my thought is way faster than the speed of light. Ahem. My mind tells me that the universe is outside of the realm of our experience. It is not in our construct to understand it. But we are not ready to believe it. It must not be true. We carry on exploring the tunnel. But is there an end where the answers lie? The light at the end of this and every other tunnel seems to be the guiding light to the next tunnel. Well, it's pointless being left behind in the darkness. We must continue following the light. Onward.

One of the first exoplanets (i.e. a planet outside our solar system) discovered that was orbiting a Sun-like star was 51 Pegasi b, about 50 light-years away. Not far at all, in relative terms that is. This discovery must have been reassuring that the process through which our solar system was created, played out elsewhere as well. It is a different matter that while a lot of us know about our solar system, very few understand it. Anyway, that planet is also referred to as Bellerophon, thanks to our compulsion of naming planets after mythical characters. It is larger than Jupiter, yet it is closer to its sun than Mercury is to our sun. A year on that planet takes only around 4 days. Also, it may be spirally going inwards into its own star with time. It has no moons and surely, no life. But, who knows! Till we discovered that planet, we had scientifically proven that planets as close to their suns as Mercury could not be large in size. But Bellerophon proved us wrong. As it orbits its way into extinction, one can only wonder what else we don't know about that neighborhood. The universe is infinite in three aspects: time, space, and mystery.

Are there any questions you have at this point? Before we get to them, we should know that there are four types of questions that exist in the cosmos:

1) Questions whose answers we know
2) Questions whose answers we do not know right now, but eventually will
3) Questions whose answers we will never know
4) Questions we are yet to ask

Question #3 must apply to the origin of life. Life is a hard-to-fathom phenomenon to originate on its own in an equally unfathomable universe. By life, I don't mean intelligent life. The creator of intelligent life is evolution. We are talking about the first single cell life with its gifts of self-replication, mutation, and being evolution-friendly. Where did that come from? It's quite possible that life may have arrived at Earth from visiting asteroids and meteorites. But how did that life originate, wherever it did? How could something living arise from something non-living? Did it just spring up from inorganic matter, out of sheer randomness and boredom? We know that randomness is ingrained in the machinery of the cosmos. But, come on! The universe cannot get away quoting randomness everywhere. In that respect, it is equally probable that this book is about to come to life and start reading your face. Sigh. I have little doubt that the process of dwelling over the origin of life, for both philosophers and scientists, is so intense that they can lose their willingness to live. Making them simply return their lives to the universe from which they received it.

Those with some amount of mysticism still living within them, must think that the formless divine mind created the universe, put some basic matter and physical laws in and let it play out. The only other thing it added at a later point was the seed of life. Our planet is one of the billions of places where the seed was sown. But if that's wrong and if there were no life elsewhere in the cosmos, it would be a fascinating topic to study on why that's the case. Also, without the Earth, there would be no intelligent and conscious observers to see and understand the universe. Thereby rendering its existence pointless. Making it non-existent. That's an interesting way of looking at it. Life has to exist in the universe because without it, the existence of the universe has no meaning. This jogs my memory to a philosophical thought experiment I had once come across: "If a tree falls in a forest and no one is around to hear it, does it make a sound?" Thought-provoking indeed. It may be the case that we are asking the wrong question, and this existential question falls in category #4 and our limited cognition abilities are simply not up to it. For now, though, the creation process and the raw material of the first life form are unknown. We have with us both the chicken and the egg but don't know which came to life first and how. The admission ticket to a life in this universe costs not knowing the answers to its working.

Man: Is mathematics indeed the language of the universe?
Wise mathematician: I do not know what language to answer that question in.

Man: I think that through mathematics, we seem to have invented something that was seemingly naturally occurring.
Wise mathematician: You may be right, as mathematics is not subjective. Our interpretations of numbers cannot differ. Our ancestors must have seen numbers as a creation of God.
Man: Can mathematics define eternity or infinity?
Wise mathematician: Let me answer that this way: In an eternal universe, the denominator for most analytical equations is infinity. Thus, those calculations have no meaning.

Disillusioning revelations are omnipresent. But, only a few develop the eyesight to see them. Allow me to tell you the story of Emmy Noether, considered by Albert Einstein and others as the most important woman in the history of mathematics. She was born in the era when women were largely not encouraged to study and allowed to hold important academic positions (what a ghastly era that must have been). I guess nobody important was ever born in the right era. In addition to fighting those odds, she carried the burden of being born in a Jewish family at a wrong period in history. In spite of all this, she fought all odds and carried on with her education and research, sometimes incognito. Supported by sane men in the academic field, she made numerous valuable contributions to the field of academics, most notable being the Noether's theorem in 1918 to help resolve some problems with Einstein's theory of gravity. You see, a bullet cannot kill an idea. This theorem came to be termed as being an invaluable contribution to theoretical physics.

What is this theorem? The theorem states that every differentiable symmetry of the action of a physical system has a corresponding conservation law. If a physical system is symmetric under rotations, its angular momentum is conserved. If it is symmetric in time, its energy is conserved. If it is symmetric in space, its momentum is conserved. What does this mean, in simpler words? Well, if you perform the same experiment now and again in ten years from now, the result will remain the same, because of factors that play out in the background. Simplifying it down further would take away the essence of it and therefore not do it justice. The understanding of it requires knowledge of symmetries and laws of conservation. Yes, science is a little hard but totally worth it. Very well, so the philosophical takeaway for me from her work is that the universe tends to intelligently correct for any changes that happen in the construct of it. It's like it's an obsessive-compulsive being who always tidies up the house once the guests leave. There is stunning precision in how the laws of conservation play out in reality. If something gains, then someone has to lose. Things are only recycled; nothing new is ever created.

This topic also brings forward the theme of constants in the cosmos. A constant is something that is unchanging with time, and universal in nature. For example, the speed of light and the gravitational constant are such constants. I sometimes mull over the idea that when everything in the universe changes or evolves with time, how can the constants be unchanging? Why should the speed of light be the same at different points in space and time? In other words, in a forever changing universe, how can we call anything a constant? As it turns out, all scientific observations to date seem to suggest that we assume constants to be forever constant for the time being, and there are 26 of them at this point. I wish that common sense were also a universal constant. It would not be prudent to talk about constants and leave the Anthropic principle out of it.

To understand the Anthropic principle, we need to know that we live in a universe that is apparently fine-tuned for life. Yes, there is obvious fine-tuning on display. We can get a sense of the degree of fine-tuning required to enable life, by looking at the values of things like the ratio of strength of electromagnetism to the strength of gravity, cosmological constant, expansion rate of universe, etc., and knowing that if these values were even slightly different, we wouldn't be here. My favorite is the supposed value of the initial entropy of the cosmos, which had to be accurate to the degree of 1 part in 10^(10^123) in order to eventually enable life. (That's 1 followed by 10^123 zeros: typing this number will take longer than the life of the universe.) While this may be a misinterpretation of mathematical physicist Roger Penrose's inferences, that's some incredible fine tuning if it's true!

The Anthropic principle says that these values in our universe are fine-tuned so as to ensure conscious life that can observe and reflect on it. Some very deep and difficult to falsify notions: conditions that are observed in the universe must allow the observer to exist. Only in a universe capable of supporting life will there be minds capable of observing and reflecting on the matter. What this means is that the universe is in some sense compelled to have us exist. This then translates to the fact that there is an intelligent design and cause behind it, thereby sparking the interest of God believers. Thus leading the non-believers to suggest that our universe is one of many (multiverse), and in other universes, the laws and constants are such that they do not support life. Ours happens to be one, completely randomly, that can support life. Hence, there is no need for an intelligent design to be behind it. Good arguments from both parties? Sometimes ambiguity is good, for it keeps both the sides interested and the agnostic audience engaged.

It is highly unlikely that a drastic event will occur in your life in the next one hour. Equally unlikely that something life-changing will happen today. Perhaps not even this year. But, it is quite certain that your life may change drastically at least once in your lifetime. As you expand the time scale, the probability of incidents happening goes from being negligible to absolutely certain. That being so, given the incomprehensible time scale of the cosmos (possibly eternal), no event is impossible. That is the power of time, making the impossible possible. Nothing is older than time. There is nothing newer than the present moment. It is now time that we discuss the most perplexing thing probably there is to understand, both scientifically and philosophically. Time.

Saint Augustine of Hippo, a Roman African philosopher who lived around 400 AD, was probably the first thinker to suggest that time may have had a beginning, and was not unchangeable and forever flowing. We still don't know if he was right. Although time is considered to be a dimension in a way, it is not a dimension like how spatial dimensions (x, y & z) are. That's because you cannot stand still in time like how you can on the road. Time forever flows in one direction only, which is from the present to the future. We experience it through changes that our brains perceive. The flow of time stops for no one. Once a moment passes, you cannot go back and change it. Not with science, not with divine interventions. On the X-axis, I can go forward and back as I please. On the Time-axis, I am stationary. The axis itself moves forward and we are bound by its order. If there is anything that is invincible, it is time. Nothing can cause it any harm or convince it to change. It listens to no one and marches on. Now then, how do we detect time? Well, coming to think of it, our brains cannot actually detect time. We rely on devices like clocks and the actions of sunlight to figure out the time. We experience time by perceiving, through our senses, things that change. In a completely stationary world, the concept of time will not have any meaning.

Theravada Buddhism says that time does not exist independently. Its existence depends on our consciousness that experiences it. Much like how psychological time is defined in the sense that it is subjective and dependent on the observer's consciousness. That's an interesting thing to reflect on. The past was an ancient era for us. The present will be an ancient era for future generations. We know that time is not the same for everyone. Objects moving at speeds close to the speed of light experience time relatively moving slower. If you were to hold this book and start running on the road right now, time would move slower for you (by an infinitesimally small amount, but still) than for the person sitting by the street and observing your absurd actions. While I am not asking you to do that, I would like to share with you a weird theory that I have about why marathon runners look younger as the years pass. You see, as they are in motion most of the time, the clock of time and thus aging slows down for them. Okay, that was pretty juvenile but still some interesting possibilities to brood about. Photons that travel at the speed of light must have no concept of time and definitely no sense of humor.

The arrow of time that we referred to is closely associated with the second law of thermodynamics, which states that as one goes forward in time, the entropy (disorder/chaos) of an isolated system can increase, but not decrease. Hence, entropy measurement is a way of distinguishing the past from the future. But it begs the question of why the early universe had such low entropy to shoot this arrow in the first place? Speaking of the early universe, how did time exactly start? Some believe it started with the Big Bang, like how space started and there was no time before that. Some suggest that space converted into time. Some, like me, believe that time is eternal and had no beginning, nor an end. But, no one can be proven right or wrong. Note that there is a conundrum in the mix, which is this: the most profound property of time is that nothing lasts forever. At the same time, time itself is probably eternal. Now, isn't that a paradox to ponder over? A thing that exists forever surely has no concept of time.

For things to be eternal, they are dependent on time being eternal in the first place. That's quite a grand assumption to make when saying things like space is eternal. I wonder how being eternal feels. Having no beginning and no end. Does the very concept of time then cease to exist? Some radical thinkers suggest that time is an illusion and reality is timeless. They say that reality is a network of events onto which we project sequences of past, present, and future. It seems that they have proof via equations in quantum gravity that exist without any need for the time factor. Well, it's quite an engrossing yet a knotty problem to have, one that we can spend an eternity on. On the subject of being eternal or not, if time is indeed not eternal and had a beginning, then that makes its origin the greatest mystery of them all. What started time? And how much time did that unknown phenomenon take to start time? A paradox we should make time to note down. Enough with these consuming paradoxes, I say! And what is this mumbo-jumbo about space turning into time? This book may as well turn into a comet.

How do you kill time on an eternally boring day? With useless thoughts like these: If our solar system has a sun called Sun and other solar systems have suns, and if our planet has a moon called Moon and other planets have moons, then by that logic our planet should be called Planet and not be given a name like Earth. Or, our sun and moon should be given a name. Imagine if we were to open that up for corporate sponsorships! That would bring some massive revenue to fund space research.

There are some massive objects out there in the universe, and by massive I mean ginormous. The biggest known star is

UY Scuti, around 1,700 times the size of the Sun. The supermassive black hole in the galaxy NGC 4889 is, wait for it, 21 billion times the mass of the Sun! I just cannot grasp the sizes of some of the things out there, and to imagine that there are so many yet to be discovered objects in the billions of galaxies floating around. I've never felt so insignificant, and yet I feel special for I can think about and make sense of all of this.

Back in the year 1969, an imaginative married couple decided to make a strange science-fiction movie, with the title 'Doppelgänger'. Its story revolved around the discovery of a new planet in our solar system. This new planet lay parallel to Earth and was exactly behind the Sun, thus always out of sight (quite possible). Also, the planet was a mirror image of Earth in every sense. So, whatever happened on Earth, also happened there simultaneously (science-fiction indeed!). When a spaceship was sent from Earth to investigate that planet, the same sequence of events happened there. Leading to some bizarre consequences. While the movie did not do well, probably on account of reduced public interest in space exploration after the Apollo 11 mission, it did bring to the table a display of our creativity in imagining the various possibilities that are possible. That story may not become true in our reality, but if all possible realities exist, then it must be true in another reality. The truth is sometimes impossible to believe. The two versions of you reading this should know this by now.

We have been collectively focused on understanding the universe scientifically, via vetted observations, for more than a century now. How much have we understood to date? What is the universe made of? Is it full of stars? Stars and stars everywhere! You may know that there are probably ten times more stars in the universe than grains of sand in all of the deserts and beaches on this planet. It may take an infinite amount of time to precisely count the number of stars in the universe. That's just to count them. Forget about getting to know them. Alas, what is the cosmos indeed made of? Well, I am sorry to tell you that we scored a sum total of 5 marks out of 100 in that examination. And that's because we have only been able to figure out 5% of its constituents, which happen to be all the known matter like stars, planets, gas, and dust. The substantial remaining portions are the elusive dark matter (~25%) and dark energy (~70%). I am happy to know that we have been humble enough to admit our ignorance and term them 'dark', learning from past mistakes of jumping to conclusions like calling the Earth flat and the center of the universe. It is just staggering to realize how much we don't know. We don't know how the first life started on Earth. We do not know for sure how the universe started. We do not know how time started. Even without answers to such most basic questions and knowing well how little we know of what we don't know, we have still chosen to award Nobel prizes to people. The understanding of the universe is equivalent to running a marathon. By the time humanity reaches extinction, we would have not even crossed the first few meters.

Man: That was a great meal. Are you the cook?
Wise cook: We enjoy the meals every day, and forever search for the cook. Will the intelligent cause that created the universe please show up and explain everything? And also bring along its creator.
Man: That's funny. Can I please have a cup of tea?
Wise cook: I cannot create a cup of tea from sheer nothing. I need the raw material, and the raw materials need a source. If I sequence it back to its origin, I'll need the universe to be created in order to get all the material to make a cup of tea. Seems like a lot of work. Can I get you some water, instead?

The universe has time and again showed us that nothing is permanent. Entities are born, exist, and then die. Our home Earth will die one day. So will the Sun. The other stars that are out there will also have the same fate. Some will fade away into the darkness silently, while some will meet their end through colossal explosions in the form of supernovae and hypernovae. The debris from such events comes together to form dust, planets, and precious elements like gold. New stars are born inside dense concentrations of interstellar gas. This cycle goes on. We seem to have incorporated this phenomenon in our local beliefs. The eastern religious concept of Samsara refers to the cyclicality of all life, matter, and existence. In its Buddhism version, it refers to the beginningless cycle of repeated birth, mundane existence and dying again. This reminds me of a joke a Brahmin told me about the three phases of life of most Indian boys born in pious south-Indian families. The three phases are:

1. Birth
2. Pray, consume prasadam and curd rice
3. Death

While various beliefs worldwide also talk about rebirth, I don't think the universe ever creates different versions of the same thing. Sure, new things may be composed of some of the same basic elements as old things, but the new things are different and unique. They have an identity of their own. At the same time, those who are not born cannot die and thus cannot become a part of this cycle. Most of those dead must regret that they did not get a chance to understand the universe better. Whatever material the universe consists of that led to this cycle of birth and death—where did that come from? And why should the universe consist of that particular type of material and not something else? Further, why should it consist of any material at all? Also, why should a universe exist in the first place? We seem to be losing our grip on reality. Let's move on. The museum of the universe has no tour guides.

"Equipped with his five senses, man explores the universe around him and calls the adventure Science"—this famous quote by one of the biggest contributors to the field of astronomy, Edwin Hubble, epitomized his approach to life. While growing up, his interests lay in sports and science, but his father wanted him to study law. He had to give in and completed his masters in that subject, as a promise to his father who had passed away. He never ended up practicing law, but instead entered the field of astronomy. Boy, are we glad he did. His clarity of thought and brilliance gave us many path-breaking discoveries including the fact that there were many more galaxies out there and not just the Milky Way, and the phenomenon of redshift increasing with distance. But his greatest discovery was in 1929 when he proved that our universe was not a static area of space, but was rather expanding rapidly. He inferred this from the observation that the redshift of galaxies was directly proportional to the distance of the galaxy from Earth. Yes, our universe is expanding. In fact, speeding up all the time. Galaxies are moving away from one another. We are moving away from some very old galaxies, whose light is yet to reach us and may never will. Is new space being created as the universe expands or is the universe matter expanding into pre-existing empty space? I don't know. What I do know is that with each such discovery, we have come to realize that the universe has turned out to be a far stranger place than we could have ever imagined it to be.

Someone once told me that I shouldn't work to accumulate assets, but rather work to obtain wisdom. I now realize that depth of thought is indeed a prized possession. It is something that is difficult to obtain and difficult to lose. Speaking of depth, how deep exactly is the cosmos? You see, I see these planets in the solar system and think that they must be falling forever because there is this force of gravity acting on everything. They have been falling for billions of years, when will they hit the floor? Okay, so my question here is quite naïve. Just like the land beneath our feet, there is a floor out there. It is the floor of the flat fabric of space-time and all planets and stars rest on this fabric. The heavier ones cause a larger dent in the fabric and gravity works its magic to get nearby celestial objects to orbit around them. There, that's how it works. This is an impossible thing to even conceive, but it was done so and proven by Albert Einstein. Depth of thought does indeed give you a great perspective on all things.

There is also this inclusion of 'time' in space-time as the fourth dimension, which I feel makes the understanding of it a little difficult. Sure, all events happen at a place in space and time. I cannot meet you at 4 PM tomorrow unless you tell me where. Nor can I meet you in the library unless you tell me when. I guess my cognitive abilities and withering brain plasticity cannot grasp the placement of time as a dimension in an equally difficult-to-grasp theory of relativity. I am willing to live with that. Oh by the way, what a magical phenomenon a 'dimension' is, isn't it? To think that there are other dimensions all around us that we cannot perceive. Physicists talk of Hilbert space as something having infinite dimensions. Straight out of a fairy tale. Our craving for the magical and extraordinary seems to be a deep-rooted trait in all of us. Although my science books in school did not have any mention of parallel dimensions or time loops. What unimaginative books those were. A failed education system, I must add. Okay, back to the three dimensions of our boring plain vanilla existence. Where were we?

Man: Can you please play some cosmos themed music?

Wise musician: Who has heard the music of the cosmos?
Man: You know, some ambient kind of meditative stuff?
Wise musician: That is what we think it sounds like. Actually, there is eerie silence out there. Apart from when stars erupt. But sound waves also need a medium to travel in and near vacuum is not the best one. Perhaps, it is in silence that the real music of the cosmos reveals itself. The theme song of the universe plays forever in the background. Some of us hear it.
Man: But most of the meditative community believes that 'Om' is the sound of the universe.
Wise musician: That is how they interpret it. Remember that once the fat lady has sung, it is down to what we choose to believe in. Music is a pattern in, and the harmony of, sounds. Close your eyes and it can take you to different points in both space and time.
Man: Your words make no sense to me.
Wise musician: The song of the universe needs no words to make sense.

The concept of color has no meaning for the man born blind. We had earlier touched upon Plato's concept of all visible things in this world being like shadows cast on a wall by the activities of their corresponding universal forms. The great thinker illustrated this via the 'Allegory of the Cave' theory where he wrote about a group of born prisoners who lived all of their lives chained in a cave, facing a blank wall. They saw shadows on the wall projected from objects passing in front of a fire behind them. The shadows were the prisoners' reality. They knew no better than this manufactured reality for they could not perceive anything else apart from the reality they saw. But, they managed to break their chains one day and discovered that reality was not what they thought it was. Are we those prisoners? Is true reality completely different from what we think it is? Do we even know where those chains are that we need to break? I guess not. All we can do is observe what surrounds us and make sense.

If we observe things around us carefully, we'll notice that the spiral shapes that galaxies naturally take can also be observed in the shapes of many things on our planet. A proof of the fact that the same laws ought to apply here, there and everywhere. Although the universe will not explicitly tell us, for it never sends us any signs. We just see signs because we are a pattern recognizing, problem-solving, and wandering advanced breed of chimps. We live on planet Earth, a planet that was formed around 4.5 billion years ago from the debris of the creation of the solar system. Our neighborhood tells us that 'the calm before a storm' phenomenon does not apply universally. On some planets, it is always calm. On some others, there is always a storm. On our home Earth, we've had pleasant life-blooming conditions mostly, and we've been able to harness the energy of various elements of the universe to sustain ourselves. Mostly heat energy from the Sun. The Sun has been benevolent enough to not send us an energy bill. We should be grateful when we can.

Actually, we should be more than grateful. We don't even own this place. The universe owns our and all planets. We are just tenants who share this with millions of other species, and we pay no rent. Have we been good tenants? I don't think so. They say that you should not play with the patience of a patient man. We have turned out to be a disjointed world. Plus, we have disturbed the ecosystem to no end. Further, we have hardly kept things in order. We do not yet have a planetary language, calendar, measurement system, a species flag, and an agreed understanding on the origin of anything. The East is fascinated with the western standard of living. The West is fascinated with eastern beliefs and practices. We are divided and chaotic. The universe isn't probably bothered, for it knows that we were created from natural elements and will dissolve back into them. We are just a temporary inconvenience for it. It has no time to acknowledge our presence or our feelings towards it. Or does it? I know all of this sounds crazy, but such is our existence.

You know that we are imperfect earthlings. But deep down, who are we really? Are we all just collections of carbon molecules, arranged in a marvelously complex way? How did we become conscious and intelligent enough to question who we are? How can an arrangement of carbon molecules try to figure out the universe as a whole? The chemicals that make me are widely available in nature and cost pretty much nothing. But the beauty lies in the proportions and the arrangement of the construct. Having raw materials is one thing. Building something brilliant with them is something else. Who is this builder who figured this out? The answer to the mystery of origins has to be encoded somewhere in the fabric of the universe. Before we attempt to decode it, we must first find it. As difficult that may sound, we need to constantly be at it.

I doubt if there is an area that has sparked as much interest of late as the hard to comprehend phenomenon called black holes. If you don't know anything about them, you've been living in a black hole yourself. Although that's not realistically possible. Black holes are reality's computer graphics. We can only marvel at what they are. Supermassive versions of them occupy the centers of galaxies, keeping everything together. None of the universal laws are applicable to them. They are the rebels who got away. The ghosts of large dead stars, they can vacuum everything near them to a state of nothingness. Often, I imagine that one day we will be able to tunnel all the waste from Earth to a nearby black hole and thus solve our waste management crisis. It's a different matter that the black hole will eat up the tunnel as well.

What stops a black hole from collapsing in on itself? Sucking itself into non-existence? Some of us who are dying to become non-existent ourselves speculate that the universe exists inside a black hole of another universe. Some others who wish to die in other galaxies propose that black holes open up wormholes providing shortcuts to other places. On a pure fantasy note, if wormholes do exist and we could pack a big telescope and get to the Virgo cluster of galaxies that are around 65 million light-years away, we could look back at Earth and see the dinosaurs roaming the lands. I am sure that some of them would have been meditating on the purpose of existence, oblivious of the incoming asteroid to end it all.

The only miracle that ever was is that the universe exists. We have mulled over how did the universe itself come into existence? How does something come into being? What events lead to celestial creation at that scale? There has to be an explanation for all of it, whether we can find it or not. Various creation stories have been adopted by different cultures in different parts of the world. From a single divine entity creating it, to a union of male & female deities, to a recycled version of something else, to just a collection of fundamental principles, to it all being a simulation—we've all pondered over this mystery and made peace with one of the theories. People seek the comfort of things making sense.

Although, the theory that we live in a computer simulation is as far-fetched as the one that says that the universe is just one of God's lucid dreams. The truth is that we don't know. The universe does not exist because of a need for goodness. Stop fooling yourself. How and why the universe began will probably forever be a mystery. Having said that, crazy theories turn out to be true and obvious explanations are proven to be false as time passes. That's our universe for you. I think that we are the result of an anomaly in the harmony that created itself. The anomaly created itself, and so did the harmony. We just happen to be brilliant products of a marvelous reality. The universe is simply an example of how a painting draws itself. Hard to believe yes, but it is more real than other things we choose to believe in.

The universe is my father, my mother, my sibling. Also, my friend. My teacher. It is you. And me. It is all of us. Like all of us, it will also have an end. The universe is all there ever was, but one day it will cease to exist. Hard to imagine, but that's how reality works. Unless, of course, concepts like eternity are not a figment of our imagination, and the universe is indeed eternal. An eternal phenomenon has no cause, no beginning, and no end. Various rich minds have proposed interesting theories on how our beloved 13.8 billion-year-old universe will eventually end. Interesting that we are so worried about the fate of the universe, which is something we will not live to witness. Here are some possible scenarios:

- Big Crunch: Gravity and other contracting forces work their magic to bring it all back to a single point (Big Bang singularity). For it to explode again, and be another event in an infinite cyclical process?
- Big Rip: The expanding universe explodes, as all matter is ripped apart into a state of nothingness.
- Big Freeze: The expanding universe keeps expanding till all stars run out of fuel, and everything dies. Into a state of never-ending darkness.
- The universe is eternal and hence has no end. Its current form may end and give rise to the next one.
- Vacuum delay: This is related to the concept of a false vacuum and has a lot of hard physics involved. Pass.

- As and when God chooses to end it (possibly after failing to understand the concept of a false vacuum).
- Our universe will collide with another universe in the multiverse.
- We are in a simulation and the real universe is elsewhere. Organisms of that real universe will eventually tell us about what the hell is going on.
- The universe does not exist. Only you, the reader of this book, exists and everything else is your mind's creation.
- Nothing exists (thus the proof for that also cannot exist).

The most profound and ultimate questions about origin and end can be asked but may never be answered. Should we care? I don't think so. In a seemingly indifferent and least bothered universe, why are we so worried and concerned all the time? We should just take it easy. You, me and the other around 40 billion human beings who have ever existed won an admission ticket to witness the wonder of existence. Let's enjoy the journey, gazing at the sights and contemplating the miracles. Join me. Perhaps somewhere along the way, we'll find what we need to look for, and then find it.

NATURE & EVOLUTION

All men should strive to learn before they die,
what they are running from, and to, and why.

– James Thurber
Further Fables for Our Time (1956)

As an evolved organism, you need to know that there is natural selection constantly at play in nature, whereby species that are better suited to their environments survive and reproduce, while others perish. Random genetic mutations across generations and thousands of years are what create subtle differences in species. All of this dates back to the first organism; thus, all species share common ancestors. As it turns out, it happened in this sequence:

Fish → Amphibians → Reptiles → Birds → Mammals

We have all descended from fish. All life originated in water. Human babies, during the embryo stage, display fish type characteristics. So, we were all a little bit of fish not so long ago. Funny then that we all need to learn swimming and it is not an inherited skill. Within the mammal family, we are descendants of apes. It's not very difficult to make that out. We look like apes, some of us more than others. Also, we like to mimic and copy others. Like apes do. The living cells within you and me are the finished product of 3.5 billion years of evolution. That's a lot of history to be walking around with. We should never underestimate the power of natural selection, which is actually just a natural process that plays out without any intelligence maneuvering it. It's astounding that blind natural selection helped blind creatures camouflage themselves in their habitats so that they could escape predators with the best of vision.

The fish from which we descended, in turn, descended from a single-celled organism that originated some billions of years ago. We do not know where that organism came from, but what we do know is that we are all related and part of the same family tree. Sometime before we were fish, we were also bacteria. That's what tickles me when I see all this nonsense about antibiotics and antibacterial soaps. We are descendants of bacteria. Without them, there wouldn't be any oxygen in our atmosphere. We owe our existence to them. They are more real than anything else we believe in. Then, why are we out to kill bacteria? Is it because we have a competitive need to kill?

Competition is ingrained in all of us. We are born with competitive spirit inside us. For the men of the human species, does the competitive spirit have its seed in the reproduction process of 250 million sperms fighting for one egg? Perhaps. Over the eons, there has been a lot of fighting among species on this planet. If you really drill down to the roots, everything is possible only because of two activities: reproduction and competition. These two created all living things on Earth. All of them. Our diverse ancestor species and we have forever fought for survival and resources, mostly food. Nobody wants to die. We all have survival instinct written in our genes. The foundation of evolution is survival instinct and adaptability. The need to survive and not die has gotten us this far. Fear of death is a motivator like no other. The possibility of annihilation brings out the best in us. We have a recent example to see this. Neanderthals were a separate species from modern humans, but similar in a lot of respects. We co-existed for a long time, but when we had to fight for the same limited resources, they ended up on the losing side and eventually went extinct forever around 25,000 years ago. Wiped out completely.

We have also fought and killed because we are all part of the food chain. Not all species eat only plants. Most of them eat other species. Nutrients and energy pass as one organism eats another. Creatures have killed each other for food and dominance since the beginning of life. Violence is in our genetic construct. That is how things work in nature. There is nothing immoral about it, although some pious humans see it as a sin. Also, why are people always after me to turn vegetarian? It is quite ironical that it took many ages of eating protein-rich meat to evolve the brains that now poke us to become vegetarian.

Wise Owl: You seem awfully stressed out. What's up?
Man: I saw two crows attacking a pigeon and trying to kill and eat it. The pigeon was defending itself bravely but was unable to fly off as it was injured. I could have saved the pigeon, but I didn't. I just walked away. I feel horrible about it now. I cannot stop thinking about the plight of that dying helpless pigeon. I feel guilty.
Wise Owl: You have done no wrong. What you witnessed was the natural food chain that is embedded in nature. It is what creates balance in the environment and sustains it. If you try to disturb it, you will end up killing thousands of organisms. You should not feel guilty. You were just a witness to the workings

of the world we live in. Just remember that nothing naturally occurring can ever be immoral. It all makes sense.

Just like most organisms on this planet, our initial days were focused only on meeting our survival needs, which were essentially safety and nourishment. Our ancestors were nomads in the wild, surrounded by predators. The law of the jungle prevailed everywhere and we used our wild instincts to meet our needs. Thus, we are all genetically attracted to wild adventure. Being wild is in our basic instinct, embedded in our very essence. The wilderness is where our species evolved. As time passed and survival became guaranteed, the brains of our ancestors told them that becoming civilized was the right thing to do. But, the wild instinct in our genes has not diminished with time. That's why children need to be taught discipline, for their natural state is wild and chaotic. One might question that has becoming civilized and creating civilizations helped us escape the survival drill? In the past, we had to venture into the wilderness to find food, braving the dangers posed by predators. We are now civilized and there are no predators out there waiting to eat us. We now venture to work every day in order to earn money for food. Stress, toxins, and unhappiness are the new predators. The survival drill still continues. Funny, isn't it?

Do you know what else is funny? The most typical attributes of the human species, namely hunting, gathering and child-rearing, are activities that most modern humans are moving away from. Anyhow, once our survival needs were met, we started focusing on our surroundings and finding ways to improve our stay here. We started making shelters, becoming organized, storing food and enjoying sunsets. Human beings evolved in forests and then destroyed them to build civilizations. As a species, I don't suppose we are very respectful that way. We all always want to live with all the comforts and facilities, don't we? We were always meant to wander, gaze, and wonder. But then we built homes and created employers. Thus, here we are.

The Hindu Vedas tell me, "There is no happiness for him who does not travel. The fortune of him who is sitting sits, it rises when he rises, it sleeps when he sleeps, it moves when he moves. Therefore, wander!" I indeed do wish to live the life of a nomad. Sell off everything. Make a camper van my home. Wander from place to place for new pastures and pleasures. Feast on local and seasonal food, experience cultures and climates, and learn and share knowledge along the way. But my logical and conditioned brain won't let me do any of this.

The most complex thing in the whole cosmos that we have come across has to be the human brain. Comprising of around 90 billion neurons, it is incredible to think that this complexity was designed by evolution. How did our brains attain this pinnacle? While our brains may not be smart enough to figure out how they work, what we know is that our brain is also linked with our consciousness, which is something it possibly developed. Our brains come preprogrammed with basic information, developed through the process of natural evolution. This basic information is in the form of our basic instincts. The ones we are born with, not the ones taught to us. Science has taught us that the first brain structure appeared around 500 million years ago, and has been evolving ever since. With time, it has increased in size, complexity, and capability. Coping with environmental and social obstacles, coupled with genetic mutations ought to have contributed to the growth of the brain. I have often deliberated over whether intelligence has actually helped organisms survive. Sure, being smarter should help to outwit competition. But, nature is full of organisms without brains, from bacteria to jellyfish, which have survived absolutely fine for hundreds of millions of years. Quite a brainy fact, that one.

Does being dumb actually help a species in survival? Thinking and planning lead to the usage of energy, thereby needing replenishment of calories for which you'll have to step outdoors to hunt and risk getting hunted. It's worth noting here that our brains use 20% of our total energy, which needs to be then restored. No brain also means no worries and no stress. As a result, there are fewer things that can kill you in the long run. Also, intelligence leads to curiosity that leads to knowledge that leads to innovation that leads to weapons that lead to extinction. So, is it a no-brainer that there is an evolutionary advantage that dumb creatures possess? Funnily enough, it requires a complex brain with high intelligence to ask and answer such a question. Such conundrums! On a non-scientific (yet not unscientific) note, is our intelligence the result of our brain wanting to understand all that our eyes can see?

Blind evolution created the eye. The eye then saw evolution in action. The eye is one of the greatest engineering marvels of reality and an area of fascination for all evolutionary biologists. Complex image-forming eyes have evolved independently several times over the last 500 million years. Eyes have come a long way from being just light sensors in fish meant to roughly detect darkness or light, to the current 500+ megapixel high-definition, high-precision camera like devices.

Nature's most amazing eyes belong to the marine creature called mantis shrimp. You should know that human eyes have three photoreceptors: red, green and blue. Birds have four photoreceptors: ultraviolet (UV) as well as red, green and blue. Butterflies have five. But none of them come close to the stunning vision of the mantis shrimp. Their sophisticated compound eyes have 16 photoreceptors, more than 10,000 lenses and can see UV, visible and polarized light! They also perceive depth and move each eye independently. It's impossible to imagine what mantis shrimps see when they open their eyes. How they detect the world around them is something we will possibly never know. It's also worth noting that these extraordinary marine creatures are known for having the world's fastest punches, carrying the same acceleration as a .22 caliber bullet. Imagine having such fighting prowess coupled with that eyesight; we can only pity their victims. Maybe evolution has been unfair when leveling the playing field.

What's that? Did I just hear you think that this is turning into a biology textbook? Well, a constant and quality flow of thoughts never made a skilled writer. But, the point is well noted. We'll just talk about the evolution of the finely tuned sense of hearing and then move on to more interesting tunes. Neither you nor I, really care about how many million years ago hearing started evolving. Let's talk about what's interesting about this subject. How did a fish gill become a human ear? Well, as jawless fish evolved to develop jaws long ago, they also formed some bones around the jaws that went on to form the primitive ear when vertebrates arrived. We should also know that the surface of the water reflects around 99.9% of the sound waves it receives. So, sound has great difficulty traveling from land to water. Likewise, as the early ear was also mostly fluid, most of the sound got reflected. It is only because of the particular arrangement of bones of the ear, thanks to tireless evolution, that sound waves managed to seep through and the gift of hearing was received and heard. While I've clearly failed to make this an interesting story no doubt, it is for us to know that all of these inferences have been made thanks to thousands of dedicated minds studying the fossil record in depth and figuring out the chain of events. I doubt if they found any novels to be of any particular interest, after having the gripping fossil records to forever keep them company.

Row, row, row your boat. Gently down the stream.
Merrily, merrily, merrily, merrily. Life is but a dream.
– *Philosophical nursery rhyme from the 1800s*

We are now moving to more interesting areas, as promised. Apparently, there seems to be nothing more interesting than virtual reality. We experience virtual reality on most nights while we are asleep, through our dreams. They seem real while we are in them. Do dreams give us any survival benefits? Or, are they just the byproducts of brains that developed intelligence? Why do we forget them within a few minutes after waking up? Some researchers, incidentally suffering from insomnia, believe that dreams have an evolutionary function, in the sense that they help us to test out scenarios that are crucial for survival. That explains why I am often being attacked or chased in my dreams. The Austrian neurologist Sigmund Freud, possibly the greatest mind to have ever studied the mind, said, "The interpretation of dreams is the royal road to a knowledge of the unconscious activities of the mind." I don't think I need to tell you that dreams help us fulfill our dark deviant desires, something which society and laws won't allow us to do when we are awake.

For us to dream, we need to sleep. Why do we humans sleep for one-third of our lives? And why do some bats sleep for 20 hours a day, while bears shut shop for months at a time? Isn't it strange that evolution would have permitted something like sleep, letting organisms lie somewhat unconscious thereby losing opportunities to reproduce, hunt and more importantly becoming completely vulnerable to passing predators? Sure, organs need rest but evolution should have found a safer and more productive way. Maybe it is still in the process of figuring that out, and one day will. As a result, we will eventually do away with sleep and figure out other ways to rest. Most of us are anyway sleeping lesser as time goes on. Note that I developed a way to sleep with my eyes open during afternoon lectures in college. So, perhaps I am already more evolved than many.

Want to think about something even more interesting? Are we born with some kind of billion-year-old collective memory to draw from, again thanks to all-answering evolution? Meaning that I am born with the stored learning that my ancestors gathered in their flawed lifetimes. Implying a kind of telepathy across generations. Extroverted people talk about having visions of previous lives. Can that be attributed to just a tidbit of the collective memory they have stored in their brains? How is that even scientifically possible? Maybe not, but we should not ridicule what we do not yet understand. Perhaps it is just that we are getting smarter as a species. Maybe evolution has sped up in the last couple of hundred years. Things that were considered very difficult to be understood or done in the past are being understood and done with great ease now. Be it math, music, sports—think about it. Children now play Beethoven's complex scores. Running a sub-4-minute mile is now a part of college runners' first checklists. If you've observed, the time taken for paradigm shifts in evolution and development to take place has also reduced as time has passed. From simple life to complex life to intelligent life to agriculture & industrial revolution to artificial intelligence—there is clear acceleration that can be noticed. Transitions have been speeding up. If evolution can't work its magic in such a short span of time, then we have to give credit for this to our collective inherited knowledge and skills. Or is something else the reason, and we are not aware of it yet? We just need to get a little smarter to figure that out. Possibly future generations will figure it out with tremendous ease. Perhaps evolution is losing patience, or we impatient beings are taking control of evolution with time.

When you think about how the food chain in nature works, you might be mistaken to think of evolution as a sadistic being, setting up killing games for its pleasure as it provides prey with a good chance to survive but also gives predators some good tools to hunt with. So, you get the irony. Such thoughts although are devoid of any logic. Hey, on the subject of logic, do you follow any superstitions or practices that have no logic or reasoning behind them? Things that we believe bring bad luck but don't know why. Some of us do. Why is that? All our brains should be wired to think logically by default, given the true reality of existence. But they are not. Why is that so? Well, we can think of evolution again. Possibly, it helped our ancestors to survive by not being logical and practical on certain occasions. Instead of employing logic to scientifically ascertain whether there was a predator in the forest before choosing to enter or leave, the chances of survival increased if one just did not enter the forest and looked for safer pastures. I guess that logic played out over generations. Clans that had only logically thinking people perished, while those with only some of such people or none of them, lived on. Maybe that's how superstitions began, by instilling fear just to be on the safer side. Natural selection evidently favored such a strategy. I can imagine children in villages being told by their parents that going into the forest after sunset would bring bad luck. The real reason would have been that there might have been hungry wild wolves roaming in those forests. Of course, we don't need to follow any such beliefs and reasoning now.

Man: My meal is never complete until I have dessert at the end to make the calorie intake value way more than what is required.
Wise waiter: That may be because you are ecologically conditioned to eat as if the next meal may never come. Your ancestors were hunting nomads in the wilderness unsure of when the next hunt would take place.
Man: Perhaps. But overeating, in this day and age, leads to obesity and ailments that will surely kill me in the long run. Evolution is supposed to help me survive.
Wise waiter: Would you like to see the day's specials or can I get you the bill?

Every part of our well-reasoned body has an evolutionary history and has taken its current form after ages of evolution. While nearly all parts of our body serve a purpose to earn their living, there are some that don't serve any purpose anymore and are just there. Probably as evolutionary relics to further drive home the point that evolution did all this. Not all of us have these additional evolutionary proofs in our bodies. Some of us know what these vestigial unnecessary parts are, with the appendix being the most well known of them all. For the benefit of most, here is a list for you to now touch and feel in your bodily museum of natural history:

- Place your arm (palm facing upwards) on the table and touch your thumb with your little last finger. If you see a raised long muscle band in the middle of your wrist, then you happen to have a vestigial muscle in your forearm. All monkeys have it, as they need more muscles in their arms to move around. Perhaps you are more closely related to monkeys than you thought you were.

- The little pink spot in the corner of our eyes has been left over from the third eyelid seen in species like fish and lizards.
- Can you move your ears without touching them? If you can, then you happen to have vestigial matter near your ear. Surround sound always for you, I guess. Our mammal relatives have it, to locate the source of distant sounds.
- Hiccups—the origins lie in fish wanting to squeeze in water through their gills.
- A small embarrassing tailbone for some unfortunate humans. Leftover from our ancestors' tails. Further proof about who our ancestors were.
- Goosebumps, which are nothing but little bumps on the skin when muscles connected to hair contract, when in cold conditions. This helps our bird and animal relatives in puffing up and keeping warm. It so happens that adrenaline hormones, which are related to emotions, are also related to the feeling of cold.
- The vestigial action of little babies clenching their wrists all the time. The origins of that lie in baby monkeys afraid of falling down from the branches.

Phew. In the wake of such compelling evidence, it is but obvious that evolution and nature made us. In some cases, they worked in conjunction with the structural aspects of the organism's body. There is this case of a goat that was born with only two legs instead of four, because of some birth defect. As the goat grew up, it learned to walk and hop like a kangaroo. This was thanks to its body that adapted and altered its skeletal structure to meet the need of the hour. It's miraculous how things naturally fall into place and do not require intelligent external intervention. It's just awesome to think that the goat and we are related, and are the engineering miracles of nature and evolution. If you believe otherwise, then this should give you enough goosebumps to make you realize that it would be a wise move to rethink your beliefs. You can even find more proof should you need, via studying our chromosomal connections with other species. It can be said of non-believers of evolution that it is sad that evolution could not make their brains evolved enough to understand that evolution created them.

Can we please have a standing ovation for evolution and nature? Nature, without having consciousness as a whole, made use of the elementary particles through blind natural evolution to give rise to the marvel of a world and conscious beings such as us. In that context, nature and evolution are our parents. They made all of us. So, as you and I appear to be then related, it would be expected that we have some affinity towards each other. We actually do. As a species with different races thrown in through evolution, we find comfort in interacting with our own kind for an evolutionary reason that we will discuss in a later chapter on values. We flock together and build communities. Species like ants and bees have taken it to a whole next level and there are thousands that live together all their lives. Is there a strong sense of security in the familiar? Yes, there is. In a large group, the predators attack the weaker ones first and thus there is an increased chance of survival for others. It is always advised to have some company instead of being alone amidst hungry apex predators. Although you might wonder on why the weaker ones join that group then. The odds of finding food are higher if you hunt together as a large group. Although the food needs to be distributed among all, so you wonder again if it actually is beneficial. Also, who is in command? There ought to be some infighting for supremacy. So, is it all worth it? Hmmm. Now that I think about it—as a species, we have no affinity for doing work that is for the greater good of all in the long term. Our affinity is for doing work that is good for self. That is perhaps because of the survival instinct that helped us evolve. Empathy for others did not get us this far.

Another reason I feel we ignorant humans get together and form all kinds of groups is that we want to know where we stand among others, both at the individual and at the group level. As competitive spirit is deep-rooted in us through evolution, we need to know how well we have done and how badly others have fared. So, we compete with one another and our group competes with other groups. Competition everywhere. An unrelated yet highly relevant example is our fascination for life on Mars. That is an example of the collective euphoria we are prone to as a species. This is a result of a combination of our innate competitive curiosity and our survival competitive need to provide opinion where others have.

It's time for disclosure. We had discussed the evolution of seeing and hearing for a reason. You see, we need to discuss a topic that nobody wants to be seen reading or heard talking: the humble topic of sex. As this is not an evolutionary biology textbook, we won't talk about how the reproductive process and organs evolved with time. My pleasure. The question we need to fantasize about is that why did evolution choose sexual reproduction (formation of new individuals from cells of two parents) as the continuity mechanism for nearly all species? Why not asexual reproduction (formation of new individuals from cells of only one parent)? Isn't sexual reproduction more complicated, in the sense that there is a pairing of chromosomes involved plus foolproof development of an embryo that needs to be mastered? I agree that we get genetic mutations sometimes through this process that help evolution, but we also get some bad mutations, which are not helpful and

lead to birth defects. Why would nature, which otherwise like a lazy student always chooses the easiest way, choose a complex way for such a critical process? I am quite intellectually aroused to quote the science writer Carl Zimmer on this subject:
"Sex is not only unnecessary, but it ought to be a recipe for evolutionary disaster. For one thing, it is an inefficient way to reproduce... And sex carries other costs as well... By all rights, any group of animals that evolves sexual reproduction should be promptly outcompeted by nonsexual ones. And yet sex reigns... Why is sex a success, despite all its disadvantages?"
Interesting thoughts. Incidentally, Carl is the only science writer to have a species of tapeworm named after him. Interestingly, tapeworms can reproduce both sexually and asexually. What a gift to possess and enjoy!

Men have always outnumbered women in the world, and I don't know what the evolutionary reason for that is. The world ought to be a calmer and happier place with fewer men and more women. Much as we would like that to be the case, all species seem to have more males than females. Can we attribute the evergreen masculine beauty attributes to this biased ratio? The male peacock, a creature of incandescent beauty, has a lot of showy feathers for a reason. With a lot of competition happening to be chosen for mating by the few available females of the species, the male organisms had to develop attractive features and skills to attract the opposite gender. Better courtship rituals than competition were required to attract a mate. Often I examine if that's why men don't seem to lose their looks as they grow older, while women do. Okay, that was a controversial statement. I never said it. As we are on the subject of gender, it would only be prudent to say that cultural aspects have also affected the way we behave, and you cannot fit evolution alone in all equations. Why do men go hunting and women gather the food? When both are equal in terms of acumen, why did women play the role of housewives for the most part of human history? Well, that's the culture we created and passed on from one generation to the next. Evolution had no role to play in creating this unfair society. We just made bad use of the brains it gave us.

Very well. We've discussed the evolution of organisms quite at length. Now, what about the evolution of the stage on which all this took place? I'm referring to the stage of our lovely and uncomplaining Earth. You know, if we were to equate the entire life of Earth to a single day, then human beings have been around for less than five seconds. And here we are trying to figure out all that happened the whole day. But the sheer poetry of this land has bowled us over. It has to be read, understood and appreciated. How did the landscape of Earth turn out the way it did? Well, nature shaped the evolution of our home and was assisted by celestial impacts from time to time. The land we stand on was created by innumerable collisions and coming together of stray objects flying through nearby space. Bit by bit, over billions of years. The entire planet initially melted, formed its core and then started cooling and solidifying. It continued to play host to visiting comets and meteorites, which carried natural raw materials. Natural processes led to the creation of an atmosphere, paving the way for life to exist. Lots of things happened on the fledgling Earth. We could spend years just going through our planet's history. Various mass extinctions, great catastrophes, and raging floods have savaged this restless land. Moving landmasses collided to form the highest mountains. In one way, our planet's magnificent landscape is a painting that drew itself. Creative inspiration is forever all around us. We should always take a moment to pause and think of the rich history of this land that gives us shelter.

Inspired humans have tried to don the role of nature in creating wonders of our own. We have created a lot of architectural marvels. Kings created many of them as a symbol of supremacy, remembrance, or worship. We can think of Machu Picchu, Great Wall of China, Taj Mahal and others. Some have been inspired by the natural wonders we see all around us. However, we stand no chance in matching wits with nature. A passing view of the Himalayan valley is enough to remind us that even with the best of skill, labor, and technology, man cannot compete with nature the sculptor.

Nature is not a content being. Nothing is ever done and dusted for it. It constantly changes the landscape, scaling new heights in perfection and beauty. The land that hosts the stunning and massive Sahara desert was once filled with numerous rivers and forests. Hey, wait a minute. Aren't humans also equally capable sculptors then, as we are also well on our way into converting numerous lush forests into massive deserts? All thanks to our intelligence. How is it that intelligent life forms like us originated on this remarkable and beautiful planet? As we have deliberated earlier, nobody knows for sure, but there are plenty of ideas that have done the rounds and will continue to do so. For example, religious creationists do not believe that current living organisms evolved from simpler organisms over time. They believe that life was created directly as we see it today i.e. human beings directly arrived in their full-bodied forms. As we scratch our heads to make sense of that theory, we know for sure that living organisms are made of cells. So, a simple approach would be for us to understand how those cells originated. The approach may be simple, but finding the answers is not. All we have are theories without convincing evidence for us to know for sure.

Life arose in water, somewhere deep down in the hydrothermal vent precipitates. The oldest known fossils that we have are of cyanobacteria, more than 3.5 billion years old! What sparked life in the first living cell? Is it just that once the

Earth cooled down and there was liquid water, that water provided an opportunity for different minerals and what not to dissolve and become nature's first experiments? And it was just this random mixing of basic raw materials that started life? If that's the case, why haven't we been able to create life from scratch in all the science laboratories of the world? What ingredient are we missing? Or is it simply impossible for life to arise from something non-living? Surely that can't be the case, as we seem to have got some proof a while ago.

Friedrich Wöhler was a German chemist in the 1800s. He was able to create something close to a miracle on a sunny afternoon in 1828. Quite accidentally, he managed to synthesize urea from ammonium cyanate. Thus, disproving that an organic substance can only be produced from living things. Also disproving the notion of 'vitalism' i.e. living objects contain a 'vital force' which is what breathes life. I can imagine how much the local churches must have hated his accidental discovery. Although, much to the liking of those local churches now is that we don't seem to have made much headway since then into fabricating important life-forming elements from inorganic matter. We have learnt that a lifeless form cannot be given life. Giving rise to another theory that life originated elsewhere and came to our planet sitting on a meteorite. How exactly did those living cells step on to that meteorite? Where and who made them? Perhaps on Mars, as we seem to be forever obsessed with that greener pasture, although it is anything but green. Maybe that meteorite was from another solar system or a different galaxy! Who knows! The origin of life is one of the biggest mysteries we can attempt to solve.

There is, of course, a simple answer that we can choose to accept, which is that God created life. Mystery solved. Why he did it, when did he do it, who is he? All of that is part of his essence and both unanswerable and unquestionable. Also, don't for once think that God is not compatible with biological evolution. The concept of theistic evolution harmonizes evolutionary logic with belief in God. That is nothing but saying that evolution is real, but it was set in motion by God. All right, we make a note of this theory as well. Full of life, albeit utterly ignorant about its origins, we march ahead.

Man: My feelings and emotions make me who I am.
Wise rock: I am lifeless. I do not have emotions.
Man: My emotions give me moments of happiness, periods of pure joy, and great satisfaction. But, they can also give me moments of terrible sorrow, uncontrollable anger, and rage. The thing that makes me human can also kill me. Why would evolution give me emotions that can put my survival at risk? Can we use logic to find an answer?
Wise rock: While it is true that logical science can explain the quantitative side of things, it cannot explain the qualitative aspects like feelings, emotions, and subjective interpretations. You have to take into consideration that you are, all said and done, an assembly of elementary particles, just like all living organisms. And through evolution, you have developed brains that have developed emotions as a necessity, just like the sense of sight and hearing. Not because of a grand design, but because of an evolutionary need.
Man: That still does not answer my question. Why would my brain think of emotions as a necessity?
Wise rock: That's because the same brain of yours thought it to be necessary to ask this question to an emotionless rock.

Some of the southern states in India celebrate the New Year festival of Ugadi. This festival is known for the festival food of 'pachadi' that is a unique combination of six different ingredients and tastes: Neem is added as the bitter taste to symbolize sadness, jaggery as sweet for happiness, tamarind as sour for disgust, raw mango as tangy for surprise, salt to symbolize fear and pepper for anger. Also, some coconut, of course, is added for good measure, as that needs to go into everything in such cuisines. Now that's an emotional and well thought of recipe. So, as can be seen, felt and tasted, emotions are woven into everything we are and do. They define us. Are emotions a result of the process of natural evolution? What evolutionary advantage, and thus a better chance of survival, would happiness, sorrow, anger, and other emotions have? If happier beings had a better chance of survival and the sad ones perished, surely by now the emotion of sorrow would be extinct. Wouldn't it?

"Love looks not with the eyes, but with the mind, and therefore is winged Cupid painted blind," said the forever-in-love English writer William Shakespeare once. Our society continues to associate love with heart and soul, and romantic tales do the rounds forever. It's true that the most important emotion is love, and love for your child in particular. Where do such strong feelings of love and attachment come from? We should be intelligent enough to know that there are evolutionary and practical reasons that can be attributed for the emotion of love. Your child carries half of your genes. So, naturally there is a natural affinity and should there be a threat to your child, your survival instincts come forward in that scenario. More importantly, in the history of any species, the survival of offspring was important for the species to continue. Parents who loved and took care of their offspring (intentionally through intelligence or through genetic mutations) thus ensured that, and those who didn't eventually lost their children to predators. Thus, species with enough offspring, thanks to parents who loved, continued while others perished. On the child's part, there was an evolutionary survival need to be taken care of in the early years of its life. That happened to be the need to escape predators and survive. This was possible only if a child was able to evoke feelings of love and care from parents and other members of the species. Those who were able to do this survived and those who weren't, perished. So, there you have it. Evolution gave us our emotions, out of the need to survive and carry the species forward. Are you sold on this?

On the subject of children, we should know that we were all premature babies, compared to most other species. We, humans, were supposed to spend a lot more time in our mothers' wombs and come out stronger and better prepared. But, thanks to evolution, our brains got so large by around 9 months of development that they could not make their way through the mother's birth canal if more time was spent in the womb. So, we arrived way earlier than we ideally should have. Thus becoming vulnerable as the body and its immunity were not fully developed as yet (thereby helping the vaccination business racket), and leading to more effort from parents in nursing and caring. In comparison, an elephant's gestation period is about 22 months and a baby elephant can start walking shortly after being born. Not so much pampering for them.

The man-made corporate business field of human resources sometimes employs the 9-box grid as an assessment tool to evaluate an employee's performance and potential. The vertical columns of the grid indicate growth potential, and the horizontal rows indicate the employee's performance assessment. This technique is used for succession planning and improving organizational productivity. I have thought about how the non-man-made fields of nature and evolution also, in a way, employ this methodology. Species in the higher-end quadrant (outstanding performance, high potential) are chosen for succession, while species in the lower-end quadrant (underperformance, low potential) are made extinct. So, I guess performance appraisals are dreaded by all species.

Sorry for that interruption. Where were we? With basic survival needs taken care of and civilizations established, the restless man now set out to improve and innovate by harnessing the resources around him. If the electrons in the atoms of his body were not at peace and forever restless, how could he be not? As he drank water to quench his thirst, I am sure his inquisitive mind must have questioned where all the water came from. He knew and we know that no life is possible without water. The brain with which you are able to make sense of this fact is made up of 70% water. Where did all the water on Earth come from?

We know that there is no feasible way of artificially making water. It involves combining hydrogen and oxygen in volatile processes that can't seem to be controlled. It is believed that some of the water was formed when Earth formed, and a large chunk arrived in the form of ice, bit by bit, over millions of years, through meteorite collisions. All of this water has been recycling endlessly through the cycle of evaporation, condensation, rainfall, and surface accumulation. The clouds don't send raindrops with parachutes. The raindrops play their role in this cycle and then transform to the next form in the cycle. The next time we take a sip of it, we shouldn't forget to think about how old the water actually is and how many such transformations it must have undergone. And when you return to the present, please do reflect on the irony that life arose inside the waters and evolved to culminate in human beings that cannot breathe and survive inside water. Although, we are full of it.

So, all the water on Earth is, in a way, as old as the planet itself. A resulting question is: how old precisely is our home?

John Lightfoot, the erstwhile Vice-Chancellor of the University of Cambridge and who died in 1675 on account of a severe cold, was of the opinion that creation happened in 3929 BC. Okay, he was so wrong but he used whatever data he had to work with. We now have much more data to work with and know that the Earth was created around 4.5 billion years ago. We don't yet know the precise birth date; otherwise, we would have loved to celebrate our home's birthday every year. Although there already happens to be an Earth Day, for other reasons I assume.

Staying with this topic, it is pertinent that you should be aware that some among us (around 40% of all Americans) believe in Young Earth Creationism (YEC), which is the belief that God created Earth and everything in it less than 10,000 years ago. Honestly, I am quite shocked that an educated American would believe in something like this. Maybe I shouldn't be, as most of the flat Earth proponents and Bigfoot witnesses are among them. In stark contrast, we under-educated yet 'over-degreed' Indians are far better off as we have no time or mind space to process things like the age of the Earth. Worth noting here that education levels in India are measured by the extent of grasp over the English language. I continue to digress, as usual, thanks to being insufficiently educated. As compared to Americans, Indians are perpetually occupied with far more important things like preparing for the next festival celebrations, figuring out the best time to leave office to escape traffic, and analyzing cricket stats to death. Should this book find an audience with a reader who happens to be a YEC advocate, American or otherwise, let me list out the data clearly. Sometimes, assumptions and lack of communication are the cause of all false beliefs. Also, the assumption that communication had already been done.

- Age of the universe

Scientific evidence: 13.8 billion years old

YEC: 10,000 years old

- Age of Earth

Scientific evidence: 4.5 billion years old
YEC: 10,000 years old

- Appearance of life on Earth

Scientific evidence: 3.5 billion years ago
YEC: 10,000 years ago

Given that I continue to have an audience with you, let me tell you about the Periodical cicadas, which are types of insects that live in North America. They are not like any ordinary insects. They are nature's amazing creations. Although they can fly, they live underground all of their lives. The only time they come out, they do it together in billions and they all emerge simultaneously. For some of the species, it is in the spring of their 13th year and for others it is in the spring of their 17th year. But they all emerge synchronously at the same time, although they are located far apart. They fly, mate, lay eggs and then disappear again for 13 or 17 years. Talk about making a vanishing act. When our clueless ancestors saw this emergence from the ground, they thought an evil plague had set upon them. I wonder if cicadas have a concept of evil. When they emerge out of the land, the first few thousands are killed and eaten by the creatures on land that are all too pleased with the surprise buffet. Once everyone is done eating and can't eat any more, the cicadas that then emerge out get a chance to mate. Do these cicadas intentionally don't come out and instead, trick others into giving up their lives for them? Let's park this question about ethics for the chapter on values.

Creatures like Periodical cicadas are just one example, among many, of the stunning diversity of lifestyles and life forms our world has to offer. With forms also come shapes and the shape we seem to find everywhere around us is the hexagon. It is nature's default shape to turn to. That is because it is the least wasteful shape. Although some of us may not know the efficiency of those shapes, all bees in nature know about them and make honeycombs in hexagonal shapes. From fruit skin to turtle shells to crystals to soap bubbles, those six-sided shapes are all around us. They are also inside us. We can find them in our DNA, sugars and amino acids. But, why hexagons? A little bit of geometry will help explain that. Makes for a quick revision in case you already know it (in which case I am delighted to have such mathematically enlightened readers): If you want to pack together cells that are identical in shape and size so that they fit perfectly, only three shapes work: equilateral triangles, squares, and hexagons. Of these three, hexagonal cells require the least total length of wall, thus making them most efficient. Makes me want to build a house with all hexagonal rooms. What's up with all these claustrophobic luxury apartments with their tiny rectangular rooms and super built-up areas?

Man: The quality of your bananas is way better than most grocery stores.
Wise fruit-seller: Thanks. I ensure that I only sell what I would eat myself.
Man: Here, taste this one. I got this from the supermarket. You'll get a fair comparison.
Wise fruit-seller: No, thanks. I eat only seasonal fruits, as nature intended it that way.

For around 90% of our planet's life, it had been home to either no life or to only microbial life. Large species like us have been around only for the last 10% and most of that time has been spent underwater. The calm seas and oceans of this world hide a vast ecosystem of life more diverse than we can imagine.

A quote by the French poet Saint-Exupéry crosses my mind at this juncture:
If you want to build a ship, don't drum up people to collect wood and don't assign them tasks and work, but rather teach them to long for the endless immensity of the sea.
It's worth noting that the endless sea took Saint-Exupéry when his plane crashed into the Mediterranean Sea in the 1940s and he was never seen again. Thereby writing the final verse in the poem of his life.

We have been able to identify and document about 250,000 of the species that live in the various oceans and seas. But that's not all of them. It is estimated that around two million marine species are yet to be documented. That's quite a lot, and therefore it is clear about how little we know of the world that is under the waters that occupy 90% of the living space on this planet. Brave divers and filmmakers have traversed the deep waters to show us the monumental beauty of that world. Along with the diversity of forms and colors, there is a great amount of mystery in those waters. I find watching such footage to be calming and therapeutic. What the fish and other water creatures must think of their earthling counterparts roaming the lands!

The waters also host the most alien-like organism to be found on the planet: the Octopus. Nothing about it is conventional and in sync with other organisms. Evolution doesn't seem to have touched these eight-armed mollusks. They seem to have landed directly from outer space, is what some claim. They have three hearts, nine brains, blue blood, and no bones. They have a brain in each of their arms. They are intelligent and have been found using tools. They are masters of camouflage and can apparently see using their skin. They can regrow severed arms. What astonishing and hard-to-believe facts these are! I can only imagine what else will we find when we discover and document more species. Our great seas and oceans sure do offer a voyage of a lifetime for those willing to dive in.

There is life in the waters. There is life on land. And then there is life in the skies. Birds and insects are bestowed with the gift of flight. On which Christmas did Santa give them this gift? The evolution of flight is one of nature's greatest feats. How did nature match its wits with gravity and win? Long ago, sometime when dinosaurs ruled the land, a creature managed to develop something like a tiny feather thanks to a genetic mutation. As generations passed, the primitive feathers developed into full-fledged wings and brief flutters of flight turned into long avian journeys. The creature with this gift visibly held an edge over others, as it could hunt prey and escape predators, with greater ease. So, this creature survived, while others died. With time, various different species of flying creatures came to be, thanks to evolution. Observant people in this field of study tell us that evolution of flight happened four times, all independent of each other, giving us large flying reptiles in the Jurassic era, to present day birds, insects, and bats.

The largest flying creatures that have ever lived were the Pterosaurs. They lived around 200 million years ago and were the first organisms after insects to evolve powered flight. While today's birds are small and light, and hence best suited for flight, the Pterosaurs were huge but still able to fly. The mechanics of their flight are yet to be fully understood. They had a wingspan of more than 33 feet! To clueless landbound prey gazing upwards, they must have looked like monsters falling from the sky. Isn't it miraculous that nature was able to make this happen, an accomplishment equivalent to getting a jumbo jet up and flying? Nature did not need to know anything about escape velocity in order to master it.

Around 10,000 species of birds glide through the skies now. Their bodies have evolved with time to become the masters of flight. Their bones turned hollow so that their bodies could become lighter. That's a good way to lose weight. With time, they have developed the ability to soar higher and faster. The endangered Ruppell's Griffon Vulture can ascend to heights beyond commercial airplanes. The Peregrine Falcon can cruise at speeds close to 400 km/hr. What amazing creatures! And boy, can they travel! Born to move, migratory birds travel great distances for better living conditions, mating, and food. Much the same reasons why humans migrate in this day and age. Birds do this without any maps or GPS apps. The Arctic tern, weighing a mere 100 grams, migrates from Greenland to Antarctica and back. That's a flying distance of more than 90,000 km. What a way to live your life; forever flying and having Earth's great landscapes to view along the way. I am happy to know that not all are forced to need passports and visas to experience new lands.

Life has evolved everywhere, from parched desert lands to pressurized ocean floors, to heat vents and icebergs. All parts of the Earth are infested with life. Some with bacteria and some with human beings. Life is like an unstoppable virus. It has also found a place underground i.e. in the soil of the Earth. It is believed that our tiny mammalian ancestors, who struggled around amongst dinosaurs, found shelter deep underground and were able to survive, both from the dinosaurs and the event that made them extinct. Sometimes I muse on the fact that if that unknown chance event had not knocked the dinosaurs out, humans would have never evolved. On a different world, there might be dinosaur type creatures ruling until the planet dies. Anyway, we digress as usual. Let's return. Doomsday prepared anxious and reality digressed humans have also gone the underground route to build bunkers in order to survive any apocalypse that may come our way. Seriously, there are so many advantages of being under the land.

You ought to know about nematode worms. They live deep down, where it is very hot and there is little oxygen. How do they do it? Well, they have the ability to become nearly dead by going into a kind of stasis for long periods of time. That's the mother of all coping mechanisms for sure. What an ability to transform yourself so that your body can cope with anything. There are supposedly ageless sadhus living in the Himalayas, possibly hundreds of years old, who have perfected this art. Did they learn it from these worms? Do these nematode worms realize how special they are? In Siberia, such worms that lay frozen dead for thousands of years managed to come out alive once the ice finally melted. If we were to blast creatures into space that could survive the unlivable conditions out there and live to reach new planets, the best candidates would be these worms. It is perhaps then fitting that we have, knowingly or not, named our most awaited discovery after them (wormholes).

There is something poetic about the wind, trees and the rain. Bring them together and you've got nature reciting its poetry. It's hard to believe that all of this natural beauty does not have an intelligent creator behind it. It assembled on its own and boy, did it get it right or what! We lead busy lives, and our limited brains cannot always appreciate the landscape that forms the backdrop. Standing here on the Earth, we take this planet for granted. It is only when we move away and see pictures of this planet taken from space, are we reminded of its beauty and splendor. Sometimes, the farther away we are from something, the better we can appreciate and understand it.

Humans share the Earth with millions of other species. It and all its natural resources belong to all of us. They are not meant for humans alone. Sometimes I feel that the unlimited supply of stray dogs at every nook and corner (third world phenomenon) is to remind us that we are not the only species living here. Also, the infinite supply of mosquitoes is also for this purpose I presume. But we choose to act like kings. We have tamed all that we could find. The only species left for us to tame is ourselves. We have chosen to domesticate a lot of species and done some evolution-type artificial selection of our own to create cute dogs we can adore, and then decided to imprison them indoors. Apartments are concentration camps for pets. They might appear happy for they know no better, but that is not how they were supposed to live. Are we completely out of tune with nature? Yes, but who cares? We have no accountability and no shame. Okay, maybe we do have some shame. Out of the multiple millions of unique species on this planet, human beings are the only ones who feel the need to wear clothes.

Somewhere, we need to accept that the scientific advancement we have made as a species comes at the cost of impacting nature. But, don't you agree that we have gone a tad too far? If we have the problem of overpopulation and unemployment at the same time, then things don't add up. Extra humans are not helping us advance as a species faster. Also, I don't think I need to tell you about the changes we have caused to the world's forests. Among other things, I am most bothered by light pollution, for it has spoilt all of my efforts of stargazing on clear nights. Light pollution also impacts hormone levels, ecological systems, bird migration, and nocturnal habitats. Let's switch off all the lights everywhere in the night, please. Let's be asleep indoor and in sync with our body clocks. Can do?

We are possibly so out of sync with nature's natural processes, because we are somewhat dumb, although we don't know it. We are told that we use only a small fraction of our brain's actual immense potential. Well, that is not by choice. We are wired that way. But why is that the case? Why this massive imperfection in our basic construct? Nature has given us a supercomputer but no manual on how to use it fully. Possibly, nature got it wrong or it may be that our brains are yet to fully evolve and we are in that difficult in-between stage. We should still know that all land, air and water bodies could never be owned by anyone. They belong to mother Earth. But we know that parents and children are known to not get along with each other sometimes. Jokes aside, it is advised that like good tenants, we keep the home in good shape, as we may leave one day, but the home will remain.

In great civilizations, there lived great men and women. There also lived great animals and trees. The Earth is a stage where characters appear, play their roles and then leave. It is estimated that between 10,000 and 100,000 species become extinct each year. Around 99 percent of all species that ever lived on this planet are estimated to have become extinct. By species, we don't mean just flesh and blood creatures. They also include plants. We tend to forget that they are alive and the fact that our existence is impossible without them.

We have dated trees and found that some of them are more than a thousand years old. One of them is more than 5,000 years old! And then there are tree systems wherein a single root gives rise to many trees, like many arms in a single torso. They are, in essence, a single organism. Utah's Fishlake National Forest hosts one such organism. There are more than 40,000 trees, spreading across more than 100 acres, all stemming from a single root. This organism has a tale to tell. It is more than 80,000 years old! And some, who believe that we cannot tell the exact age by studying its tree rings, believe that this organism's actual age is close to a staggering one million years. This organism must have seen a lot in its lifetime.

Man: What are you made of?
Wise tree: Like you, I am mostly empty. It's remarkable to note how much empty space exists in our very own bodies. Yet, we keep ourselves together as one being.
Man: In terms of overall mass, yes we are mostly empty. But I have a lot of cells inside me. Do you also have cells?
Wise tree: Yes, I do. Lots of them, like you. No life is possible without basic living cells forming the foundation. Cells play their role and die. Then new cells are formed. Cells that don't die cause cancer. It's important for cells to die on time.
Man: I know that trees, in general, live for a very long time. Some trees we have found are thousands of years old. Do you fear death?
Wise tree: We exist for as long as nature allows us to, and then we die. Sometimes, you humans kill us midway. But that's okay. We hold no grudge. In a land with limited space, you must first destroy in order to create.
Man: Your aged wisdom is unparalleled. Please tell me what you have understood about the workings of nature.
Wise tree: Nature is a puzzle to be solved, with clues scattered all over the place. But, I am afraid I cannot move. You need to explore and find the clues, and eventually the truth.
Man: Say no more. I am on my way and will return one day.
Wise tree: Remember that there are more things that you cannot control than those that you can.

Does the truth hide from us because evolution has hidden it in some corner that can never be found? As we have seen, the truth about existence is not critical for existing. Beings can lead safer and longer lives if they do not indulge themselves in such unsolvable mysteries. By contemplating such matters, they are wasting energy and losing focus from meeting survival needs. If it doesn't help you survive and reproduce, evolution will keep you away from it. Like a caring parent taking care of children. Hmmm. We'll reflect on nature's philosophies on some other morning.

On the morning of 26th December 2004, the greatest natural disaster struck the planet when a devastating earthquake hit the Indian Ocean. The entire planet shook. Tsunamis that measured 30 feet high wreaked havoc killing close to 230,000 people and countless animals and plants. The good and the bad cannot exist independently. We have come to realize that there is a very bad side to everything that is mostly good. Including nature. It is not accountable to us. It will create earthquakes and tsunamis when it is time. It is not seeking a tag of being trustworthy. It begs no acceptance. It is different from others in the sense that it is completely indifferent. It does not know that it is being bad and we cannot ask it to change its mind. It is like the cosmos we live in. Nothing here is by design. Some consequences are a result of our actions. Everything else is random and completely by chance. When the time comes for it to rain, it rains. Irrespective of whether you've made any outdoor plans for that day. Those don't matter to the rain. We cannot control the natural environment around us. We can only take shelter.

Things did not start falling down once we discovered gravity. They always did. The laws of nature are not dependent on being figured out. They do not seek understanding and appreciation. We should be like them. There is a lot to learn from nature. We have learnt from it that the process that always runs in the background is the process of natural evolution. To that, we have added the process of technological evolution, which we always seem to be in the midst of. Our wandering ancestors must have seen fire as a technology marvel. We have come a long way since. Hero of Alexandria, an engineer in

Roman Egypt in the first century AD, invented the 'eolipile' by using a heated water container to move a turbine and, as a consequence, set the ball rolling on the process that would lead to the revolutionary invention of the steam engine. Steam engines gave us the power of mobility. We could now get goods and ourselves to places faster and more efficiently. The drivers of the first steam engines may have seen it as an advancement over horse carriages, but through its invention, the steam engine opened our eyes to how we could make use of the energy available all around us. Since then, we've figured out ways to channel the power of nature to power our homes and cars. Thanks to the previous chapter of this book, we know a lot of the potential of the energy in the cosmos. From the Sun to water and wind, to fusion and fission, the energy reserves run deep and are waiting to be tapped.

Man: I am trying to think that which invention of ours has been the most important in history?
Wise carpenter: We cannot nail it down to any specific invention. Every era has given us something new to work with, and one thing has led to another. Every new invention has its roots in the earlier ones.
Man: Surely, computers have to be our greatest work.
Wise carpenter: Not really. Computers and automation owe their existence to the inventions of machines in the times gone by, which in turn are thanks to the invention of basic tools like the pulley, long ago.
Man: Right. There is also this talk about virtual realities coming up. I imagine if something like that is even possible?
Wise carpenter: The fact that we can, possibly in the future, create virtual realities with conscious beings is truly astonishing, for we have evolved from single-celled organisms that did not have a brain.

Man: I think that's pure science fiction. But then, till only a few decades back, the notion of video chatting with someone using a handheld device was an item of science fiction fantasy. Today, it is an everyday trivial affair. Today's sci-fi movies are full of stunning technology.
Wise carpenter: As a matter of fact, science fiction literature has contributed immensely in stimulating the minds of inventors. History has proven that today's science fiction is tomorrow's real science. I can't even imagine what impossible fictional notion today will become commonplace tomorrow. So long then, let me return to my work. Got a chair to make.

As I type this sentence on a computer, I cannot help but think that the credit for machines still needs to be given to natural evolution. As it created the human brain that then created the computer and everything else in the technological evolution. I cannot help but repeat that nature, without having consciousness, made use of the elementary particles through blind natural evolution to give rise to conscious beings such as us. And now we, with all our consciousness and intelligence, have used the same elementary particles and artificial evolution to give rise to machines devoid of any consciousness. Although some of us worry that those machines will develop intelligence of their own one day. As a species that's now completely dependent on technology, perhaps technology can solve that problem for us.

One can be skilled, but not be intelligent. In fact, a combination of both is quite uncommon. We have used our skill to make this state-of-the-art world, but have we been intelligent enough to ensure its safety? Weapons of mass destruction are a byproduct of the technological evolution. They have been used in wars in the past, and now pose a threat to annihilate the planet should a political leader lose his head one day. Our quest has barely begun, and we are already exhausted and on the verge of blowing everything up. Perhaps there were other planets out there in the universe that developed intelligent beings like us who then blew up the planet. I hope we don't head that way. But, our track record is not great. We have dropped bombs in the past. Our dumbness and our brilliance are inseparable.

Human beings are the finished product of billions of years of evolution. It would be a waste of colossal proportions should we all just be wiped out in a nuclear bomb disaster. Like any corporate business continuity plan, we need to be ready to recover when disaster strikes. One of the solutions therein lies in setting up colonies elsewhere. Thus, should Earth cease to exist, we will have human life available on other lands to take humanity forward. And, like Noah's ark, it would make sense to take other species with us to other lands as well so that the ecosystem can be recreated. Also, as the most advanced species, it is our moral obligation to take care of others. We should not be hesitant in getting started on this exercise. Nature is sending us enough signs already, albeit not directly. It's like nature is a bottled up passive aggressive creature. The sea never refuses any waste; the air accepts all exhaust fumes. They come back to us with a response later. The response has now come in. With overpopulation, depleting resources and global warming, our home is getting worn out. Venus is an example for us on what a runaway greenhouse effect can do. Mars shows us what a lifeless land devoid of natural resources looks like.

We see what we choose to see. Our ancient ancestors were gatherers of essential resources required for sustenance. We are creators and gatherers of clutter, for which real estate is running out. Our growing inability to be in equilibrium with the nature around us is additional evidence that we are ready to venture out to new cosmic lands. The bird in the nest is now ready to fly.

As we wrap up this chapter, we know that evolution will continue to happen in the background till you finish the book and forever thereafter, it makes sense for me to put forth a couple of maddening utterances:

- Why doesn't evolution apply everywhere? Why don't stars that have better chances of survival (via size, etc.) get assisted by it? Why isn't there grand natural selection going on at the biggest scale?
- As we fail to locate the first cause that set in motion the cause and effect cycle, is it just possible that the first cause evolved from an inferior set of causes that did not have any effects?

Why isn't everything impacted, aided, or eradicated in the cosmos by evolution? I guess that's further proof that there is no intelligent and consistent design in the making of our cosmos. Things just happened. They played out as per the governing laws. Mostly randomly, evolving and changing with time. Things somehow fell into place. What's now left is for us to find ours.

MIND & CONSCIOUSNESS

It is good to be a cynic, it is better to be a contented cat, and it is best not to exist at all.

– H.P. Lovecraft
Collected Essays 5: Philosophy, Autobiography and Miscellany (2005)

If you can read this, you happen to be a conscious being. All conscious beings are aware and can experience reality, including their thoughts, feelings, memories, and the environment around them. It is only through our conscious experience that we are able to perceive and make sense of what we experience. It is what turns your brain into a mind, and at the same time, so baffling that we are yet to understand what it is and where it comes from.

Is consciousness or awareness just another feature that our brains developed as a result of natural evolution? In a primitive land filled with predators, did beings with less consciousness have a lesser chance of survival than ones with more of it? Or did having more consciousness give you an edge over others in being chosen for reproduction? These are some startling questions requiring some supreme awareness in order to be answered. Many have attempted to answer them, but none with conviction. The American philosopher Daniel Dennett is of the opinion that our brain is like a smartphone and consciousness is the screen. That is to say that consciousness is in no way influencing our brain; it's only how we interface with it. A phone can function very well without a screen; it's just that it wouldn't be usable. That's quite an interesting analogy, though not applicable for the pre-smartphone era. Ahem.

What does having less or more consciousness even mean? I do not know. Are we kidding ourselves when we try to define consciousness as some neurons or chemicals inside our brains? Would it be better to define them like an unexplained soul that takes form when you are born? Is it more fundamental than we think it is? We know that everything that we do gets stored in our memory. Can we recall any part of it as required at any given moment? I am just giving you a set of questions without any structure and that's because none of us know much about this subject. For that matter, none of us know much of the human brain as well. The MRI scanners did not become widely used till around the 1980s, so we haven't had much time to study the indecipherable brain as yet. Whereas, we have been trying to explain consciousness scientifically and philosophically since the 1600s.

We know that our brains are a part of our bodies and are connected with all the organs. Does the same hold good for consciousness? Is it connected with the body or separate from it? Our minds have now evolved enough to study their own philosophy, which leads to problems like the mind-body problem, in the sense that are they separate or one thing? Yes, there are two schools of thought on this one. Dualists believe that the mind is an independently existing substance, while monists don't believe so. I think the basic problem why people are divided on this is because no one knows how consciousness can arise out of a biological and evolution-built organ like the brain. It's like breathing life into a rock. How is it possible? Or is it an illusion and there is actually something else which is real? Oh, how the dreamer in me wishes that it turns out to be something that transcends the dimensions of space and time. What a mind-bending story that would be!

Our minds have a soft corner for stories. Let me tell you one. It is the romantic tale of a man who refused to study and learn from books and instead chose to travel to read the great book of the world. René Descartes was a French mathematician in the 1600s. Like most mathematicians of that era, he was also a scientist as people were less insecure about their subject matter areas then. He was also a philosopher, and not just any other philosopher. He is considered as the father of modern philosophy and the reason why people took to understanding the world by the use of reason. Like many great minds, he was self-absorbed and not equipped with the best of socializing skills. That was probably good, for it allowed him more time to argue the existence of God, come up with concepts like mind-body dualism, and find our place in nature. All of which are still the pursuits of our efforts. He is best known for his phrase "Cogito, ergo sum" which translates to "I think, therefore I am." By that, he meant that you couldn't question the existence of your mind because questioning indicated thinking, and thinking indicated existing. In other words, the very act of doubting that you exist is proof that your mind exists, for it is able to doubt. Now that's some deep thought, don't you think? You should also know that alongside all of this, he also laid the groundwork for analytic geometry. He believed that science would be the main source of true wisdom, and we cannot just rely on our senses for everything. What a great mind he must have had, so much so that his skull was stolen from his grave!

His ideas may get corrected and be forgotten with time, but the man must not be.

Some inspired philosophers have built on Descartes' insights to stand up and say that in the absence of conscious beings that can observe, perceive and understand it, the existence of the universe has no meaning. Some of them have repeated that the fact that you can question your existence is proof that you exist. Additionally, they have added to that argument the notion that everything around you is just a figment of your imagination. Meaning that only 'you' exist. It would be interesting to gather such philosophers in a room and have them debate about who among them actually exists. Joking apart, the philosophy of the mind and consciousness is a topic so interesting that it can just consume you. At the same time, existentialism is not to be thought of as a subject of despair, but rather something that will help you get out of a phase of despair.

Seriously, what makes a being conscious? Isn't my brain just an organ that developed through evolution to keep me alive? It's meant to send a signal to my hand to withdraw when I try to touch fire. When and how did the brain develop consciousness to question its own existence? As we discussed in the previous chapter, there seems to be no evolutionary survival advantage by having consciousness. In fact, it reduces your chances of survival. Many philosophers have committed suicide, after entering depression thanks to studying such profound topics. It's perhaps good that plants and trees supposedly don't have consciousness. Also, dogs. Are dogs self-aware? I think they may be. I feel that stray dogs get suicidal and so some can be found sleeping on side of busy roads, knowing very well that they could easily be run over and killed. I do think that's the case. How are you going to prove me wrong? Those dogs can't testify. Till the time we can read minds, it would be wrong on our part to rule out consciousness in other living things.

Man: I was born with basic survival instincts inbuilt. My heart beats outside of my control. My actions are conditioned and dictated by the society I live in. My body asks for fuel when it needs it. So, what role do I actually play in my own life, apart from being a supervisor of my material body?
Psychiatrist: Everything cannot be known. Our intellect can take us only so far. We should make peace with that. I am going to prescribe some tests and medicines, which should help you with that. Any allergies?
Man: I have a dust allergy. Sometimes I wonder on how can we be allergic to dust when the universe is full of it?
Psychiatrist: Are you sleeping well in the nights?
Man: Man is cursed with not having dreamless sleep. The mind has no break. On these dark aimless roads, we walk blindfolded. Trying to understand the concept of a destination.
Psychiatrist: You're quite the philosopher I see. Take the medicines and things should be all right.
Man: What happens if I don't take these sedatives?
Psychiatrist: You're not going to die. It's just that your thoughts will continue to keep you restless.
Man: When the restless adventurer becomes depressed, you know that something is wrong with the world he lives in.
Psychiatrist: See me in two weeks, please.

Things that happen around you are most often not in your control. You may feel that you have everything sorted out, but something beyond your control will happen to disrupt it all. Most often, even your thoughts are not in your control. You are thinking about an apple now. See, how easy it was for me to make you think of an apple. On a more serious note, nobody wants to intentionally have negative thoughts, right? But they do indeed manage to fill your mind some or most of the time. You don't know the next thought that you are going to think about right now. It is not under your control and will just appear. So, are you just an observer of your thoughts, exercising whatever little control you have from time to time? Our cognitive abilities are good but limited. There will always be things that will be beyond our understanding. Some have made peace with that.

Sometimes, the most beautiful poems and abstract thoughts come out of the most logical thinking minds. The poem Metamorphoses, written by the Roman poet Publius Ovidius Naso and containing a mere 12,000 lines, recited the story of a sculptor who fell in love with a statue he had carved. He prayed to the Gods for a wife who would be, as he prayed, "the living likeness of my ivory girl." That ivory statue came to life one day, thanks to divine blessings, and they got married. That story makes us question if non-living things can have consciousness. Surely not, for they have no life and no brains. A stone does not care if you step on it or not, nor does it question why it exists. However, can consciousness exist in a dormant state? If yes, is that what lifeless objects have? Anything that makes you believe that it exists must be conscious and intelligent. Also, aren't we nowadays talking about computers and machines becoming conscious and aware, and taking over the planet. What's all that about? Or is the concept of living and non-living just our perception, and we are all both dead and alive, and also neither of them—all at the same time. I seem to be losing it. Oh, how much I crave to have the unchanging peace and carefreeness of that lifeless rock lying on the ground.

My full-of-life brain can only process what my eyes can see, and my eyes see only what my brain is prepared to understand. Maybe there is more out there that I just can't perceive. There must be more clues and information out there that could lead us closer to understanding consciousness and the eventual truth. If we created a one-dimensional simulated world within ours, creatures in that world would just experience dots and the passage of time. They would be oblivious of the things happening in the three-dimensional space, although it would be all around them. Is something similar happening with us? We continue to forget that our cognitive abilities are limited, and there is so much out there that we can't even perceive. What is the difference between actual reality and mere appearance? The brain that sees reality is also the one that hallucinates. How do you know which is which, and if your mind is playing tricks on you? The puzzling plot thickens.

Making entry into the waiting list of things to be understood is the concept of an altered state of consciousness (ASC). This happens to be the state that you supposedly enter when in deep meditation or when someone 'hypnotizes' you or when you are under the influence of forbidden narcotics. It is an altered state of awareness that is away from true reality, but all so real at the same time. I am yet to experience this state. Although if daydreaming is a milder form of ASC, then yes, I have been there. Many people in the world have experienced it fully, and attribute enlightenment and various great inventions to such altered states of awareness. Perhaps moving the mind away from our everyday perception helps it to peep into what we cannot perceive. Some out-of-work musicians have also claimed that music (in harmony with narcotics no doubt) has helped them enter into all kinds of altered states. Indeed, music can lift your spirits. I don't know how and why those arrangements of noises can move us the way they do, truly making us happy to be alive.

Here's another true story from the past. Back in the 1800s, somewhere in France lived a woman by the name Irène. She was about 20 and was taking care of her mother who was terminally ill. Her mother was all Irène had in her not-so-happy existence. The illness eventually took her mother's life and left Irène totally shattered. The trauma completely consumed her, leaving her depressed and without the will to live. Her mother's last moments before death filled Irène's mind all the time, leaving her incoherent and detached from everything. Her friends arranged for a psychologist to help out. The psychologist Pierre Janet was an upcoming expert in this field and started the difficult treatment. He made Irène restore her memories, via hypnosis and while she was conscious, and made her narrate the chain of events. It was clear that her brain had not realized her mother's death, although she had witnessed everything. Through his sessions, Pierre was able to make Irène realize that her mother had passed away for good. Subsequently, her mental condition started improving. The reason I have dug up and narrated this story is to throw light on the peculiar ways of our minds. The mind can dissociate itself from reality thereby refusing to realize what is true and what is not. In a way, this is also a form of ASC. This working of the brain will help us in the later chapter on people's belief in God and religion. For now, we need to talk about this story's psychologist, Pierre Janet, a little more. Through studying cases likes this, he made us understand the realm of the subconscious.

Pierre Janet coined the term subconscious and defined it as an awareness layer that was present beneath the regular functions of a conscious mind. I know that consciousness is itself a baffling concept to understand, and now I am taxing your minds to explore the subconscious. But without understanding it, we cannot attempt to understand how our amazing minds work. I am sure some of this you subconsciously already know!

My subconscious is the part of my mind that I am not fully aware of, but plays a part in my actions when required. I like to think of it as a backup generator that is always on and kicks in when there are power outages. Most of my routine driving on roads is done subconsciously, for my conscious primary aware mind is preoccupied with other matters. You can think of it as the autopilot mode of the mind. It is also what wakes me up just before my alarm is about to ring. My subconscious is formed on the basis of what I experience and learn through my conscious primary mind. In a way, all of that information gets processed and stored in my subconscious, to be retrieved and used as and when called upon. Probably that's why so many professors are absent-minded. The vast amount of knowledge they accumulate by leading nerdy lives gives them a super powerful subconscious mind, whereby the primary layers of their minds can then always be elsewhere and fantasize about topics like how life would have turned out had they chosen sports as a career.

How did this additional layer of the subconscious develop in our brains? Although by listening to my answer you are assuming that I understand where consciousness came from, which honestly I don't, I will make an attempt to sound wise and answer it anyway. Assuming that the brain developed consciousness as a combination of some neurons and chemicals in its evolutionary journey, the subconscious ought to have come into existence via genetic mutations, and helped organisms use mind space to deal with only important real-time survival problems at hand. Everything else that was not important was passed on to the subconscious part, to be used if and when required. This theory is simple and makes sense, but there is no way to prove it. Also, on this subject, does my heart beat as instructed by the subconscious part of my brain because, for sure, I am not consciously thinking about my heartbeat every second. It happens automatically like breathing. So, are such processes managed by the subconscious or are they not controlled by the brain at all, and are just a default function of those organs? Have I made it more confusing? You bet I have.

You know, there is an agent in my brain that keeps sifting through the archives of my mind and retrieving stuff I can get stressed and confused about. I've just recollected that there is something called the unconscious. Please note that the subconscious is not to be confused with the unconscious. Yes, there is one more layer in the mind i.e. the layer of the unconscious part. This part contains traumatic, repressed or socially unacceptable thoughts that can never be deliberately brought into the primary conscious layer awareness. They can only be retrieved if some unique circumstances arise. This layer is very much part of the mind and as a result, impacts the other two layers in some way. So, we have three layers of awareness in our minds. Blame Sigmund Freud and others for turning one problem into three.

We had earlier touched upon the probable fact that we are all born with some kind of billion-year-old collective memory to draw from. The Swiss psychoanalyst Carl Jung had coined the term collective unconscious explaining that a part of our unconscious mind consists of an inherited collection of experiences, shared with the rest of the species. Wow! Here's an example: An experiment was conducted wherein 100 young children were made to interact with a rabbit and a snake for the first time. It was observed that most of the children attempted to interact with the rabbit by touching and talking. None of them were willing to touch the snake, although they had never seen a snake before. The inference made was that fear of snakes was part of their collective unconscious, inherited through many generations of dangerous encounters of humans with snakes. Isn't that amazing? In addition to common evolution, how much of the collective unconscious that we share has got to do with the fact that you and I happen to be composed of the same elements that were formed in stars that died millions, perhaps billions of years ago? Our family tree actually began there. I have to be honest with you. I am deeply moved by the fact that you and I have a connection at such a fundamental level. Although we have probably never met or discussed this, we are more related than we can imagine. We should definitely plan to meet sometime soon.

Are you okay meeting a stranger? What if he turns out to be a psychopath i.e. a person without a sense of right & wrong, and capable of doing anything, with no fear of the consequences? Do people become psychopaths by choice and intent? No. It is all due to the workings of the mind. There are physical differences in the brains of such people, which make them different. They are wired differently. They have no fear, no sense of guilt, find blood calming and are very good at acting. They can act like perfectly normal beings. Does a psychopath realize that he is one? That he is a threat to others. Can he understand himself?

Can you truly fully understand yourself? Who you are and what are your limitations? I don't think you can. Nobody can. You are not one straightforward person. You are a complex work of an infinite chain of events that have played out since the birth of the cosmos. The second lesson in understanding yourself is that it is not about you at all. You are a minuscule part of a grander picture, but at the same time, the picture is incomplete without you. Third, always remember that reality is subjective. We all perceive reality differently. The thoughts that cross your mind when you see a cat crossing the road are different from the thoughts that cross my mind. That makes each one of us unique and special. Yet connected, as we have learnt.

Somewhere in Greece, sometime around 500 BC, along came the philosopher Anaxagoras, introducing the concept of Nous or the cosmic mind. This is defined as a universal mind or universal consciousness that contains all knowledge and is the essence of all being. It is from which everything is thought and given form. It is the absolute, and hence beyond all attributes. It's everywhere, all the time, all-powerful and all knowing. That to me is the very definition of God in religions like Hinduism. It's like a super consciousness, which sees and experiences everything. Thanks to our limitless imagination, I feel that we have extrapolated what we experience in our lives, to the grandest scale. We will always be better philosophers than scientists. Although we have a sufficient amount of science in our minds that tells us that it is downright improbable that the universe has a mind of its own as a whole with consciousness. It's actually not improbable, but fully impossible. But in 500 BC, we did not know much about the workings of the cosmos, so such flights of fancy were worth embarking on.

The past was the present when it happened in the past, and the future will be the present when it happens in the future. What's always real now is only the present. Everything else is sculpted by our minds. Memories of the past and hopes for the future—both exist in our minds simultaneously. Discussing such topics can make people sound crazy. But do remember that there is nothing wrong with anyone. We are all doing what our minds tell us is the right thing to do at any given moment. As you and I know, we interpret the same thing in various ways, all of which appear right from the point of view it is looked at. Likewise, all actions can be justified in our minds. We can be at peace irrespective of the action and the outcome. It's just a play of our minds. Our minds make us both an ignorant fool and a well-read wise man, and all the states that exist in-between. Without our minds and their gift of thoughts, there is no universe for us.

Man: I always wake up feeling quite groggy and lethargic.
Wise lunatic: Your mind defines you, and your body does not. Anyway, why so groggy my friend?
Man: I am not your friend.
Wise lunatic: We all carry the same physical features and share a collective unconscious. Yet, we never want to be on the same team. Please tell me what bothers you.
Man: I am yet to understand the meaning of everything, and it is nowhere in sight.
Wise lunatic: You have one small tool to figure it all out. It's small in size but massive in capacity. That would be your brain, which once conscious becomes your mind.
Man: You say that but who can understand or define consciousness. Or even imagine it.
Wise lunatic: Only the fool imagines the unreal.
Man: Our minds are just products of evolution. In sync with physical laws and carrying out their evolutionary purpose.
Wise lunatic: They are so much more than that. If you close your eyes and imagine, you can be in any place and time that you wish to be in. That is the immense power of the mind. Your mind is a time travel machine that you were born with and will always have. It doesn't need to comply with any laws of physics.
Man: What reality do you live in, my friend?
Wise lunatic: Reality is subjective. Don't lose the reality of that statement.
Man: Who let you out of the psychiatric hospital?
Wise lunatic: You cannot be separated from something that you were not attached with. Anyway, I don't know for sure if I exist. Good luck, in any case.

The mostly luckless field of human resources sometimes employs psychometric profiling as an assessment tool when hiring for certain roles. These profiling tests are aimed at measuring mental capabilities including personality and attitude in an objective and unbiased manner. They are supposedly reliable indicators of actual future job performance. I have often thought about how can we use the measurement of mental faculties on one random day and generalize to label it as the unchanging mind of the applicant. Surely, we haven't understood the workings of the mind well enough to claim that we have mastered the assessment of it. In an evolving reality, things change with time. And intelligence, or the lack of it, is never consistent.

Existence brings happiness and sorrow in equal measure. More sorrow than happiness for many. It brings problems that need to be resolved, expectations that need to be met and illnesses that need to be cured. So, why do we go through the bother of existing? As we have noted, if conscious and aware beings like us choose to no longer exist, then the existence of the cosmos loses meaning. Anything that is devoid of conscious observers may exist in reality but is as good as being non-existent. We need to exist for the word existence to exist, and our existence is nothing but the existence of our minds. Without our minds and their consciousness, our imperfect symmetrical bodies have no shape or form. Furthermore, we know that the presence of observers changes the very nature of reality. By your inquiry and reading of this book, you make light photons bounce off the pages and hit your eyes. You thus change this book, though, in a very insignificant way, but you do. And you do that to everything around you. This is called the Observer effect in physics. By the way, thank you for reading this horribly structured, flowless and typo-ridden book. I know that this book exists, for I wrote it. But without any readers whatsoever, does the existence of this book have any actual meaning or purpose? Some might argue that they have found no meaning even upon reading it.

The presence of observers or the absence of them brings out of the darkness the enigma of quantum superposition. There's no better way to understand that than by thinking of the Schrödinger's cat thought experiment: A hypothetical cat sits in a closed box with some harmful stuff, which if triggered, can kill the clueless cat in an instant. There exists a random probability of that harmful stuff being triggered. So, when we open the box, the cat can be dead or alive. Thus, till the box is opened, the cat is simultaneously both dead and alive, for it's only an observer who sees and determines whether the cat is dead or alive. In the forever absence of an observer, the cat will forever be dead and alive at the same time. In the context of this book, think of it this way: It is possible that the last ten pages of this book are blank, or contain Sanskrit verses or photos of alien evidence. All these three possibilities exist simultaneously, till an observer observes those pages. Got it?

Your mind is the observer and it makes reality real. Your mind has thought of uncountable thoughts in your existence so far, continues to have thoughts at this moment and will continue to do so till you exist. We had earlier thought about the matter that we are just observers of our thoughts, exercising whatever little control we have from time to time. We are guided and motivated by our thoughts because we have time-bound existence in this world. You see, we become passionate about things of interest when there is a deadline, external or self-imposed, associated with them. That's because we enjoy the satisfaction gained from doing the job well and on time. But, what if you were immortal and had all the time in the world? Would you still be passionate about those things? Example: The traveler stranded on a moving boat in the ocean looks forward to seeing land one day. Along with that hope, he has his thoughts for company as he stares at the waters. How well and long will he survive if the possibility of reaching land does not exist and all he has are his thoughts, the endless ocean, and starry sky to gaze at? Will he commit suicide and die?

We experience everything in the world through our senses and the tools developed with the help of our minds. But we will all die one day. What happens when we die? Does our consciousness live on in the form of an everlasting soul? What happens to all your plans and your feelings of love, hatred, grief, ecstasy, and loneliness? Does all of it just end with your mind ceasing to be alive and aware? Death is anticipated by all and awaited by some so that they can get to know these answers.

Here's what we know so far: death is inevitable. It is the natural progression of life. Your life will end completely when you die. By completely, I mean all of it. Your existence just ends. No aspect of you will continue to exist. That includes things like consciousness. You don't die with someone and embark on a journey in the afterlife. You die alone. There is no afterlife. I fully endorse this view. But, there is no way to prove it. No one dead has testified as yet.

The concept of an afterlife has mesmerized the minds of most civilizations. The Egyptians built pyramids to preserve corpses for the journey beyond. However, no one came back from the journey to thank them. Why do we persist in this effort to live beyond death through eternal consciousness and entry into heaven? Without a trace of evidence of such a thing, why does humanity continue to have faith in the concept of souls? Considering the billions of creatures that have lived and died on our planet, how many souls with consciousness must be floating around? What does a soul do all day? Where does it live? Do souls interact with one another? Do they believe in a soul afterlife as well?

In the Japanese religion called Shinto, people believe that dead people go to a place called Yomi, which is a dark underground realm that has a river separating the living from those dead. The various religions of the world have their own theories on what happens after death. Some believe in souls while some in reincarnation. Some think that the rewards and punishments for sins and good deeds done in lives come after death. Some bury the bodies while some burn them. I have come to understand that all religions tend to believe in the stream of consciousness carrying on, in some form or the other, after death. But I don't agree with those beliefs. Once dead, nothing comes back. No flowers and no people. There are new flowers and new people that take their place. In the grand scheme of things, we are born, live for a few seconds and then die. There is no reincarnation and no everlasting consciousness. It's all about those few seconds of ride time that we have to take the ride and learn, wonder, experience, and reason. Then it's over. There is no eternal soul. If anything, our brain's construction of consciousness is our soul. When the mind stops working and dies, everything associated with it also stops. I would love to know, in my lifetime, how death must feel. That state of finally being at peace with everything and ready to transcend into non-existence. But for that, I would need to die.

What good is my life without death? If I were immortal, would my existence have any meaning? Knowing that our days are numbered and there are things to be done and learnt is probably what gives life meaning. It's what drives me to accomplish things, not procrastinate, write this book and try to understand the cosmos as much as I can. Without death, life would have no meaning. We would procrastinate forever. Life and death are the yin and yang of existence. One is nothing without the other. They are two sides of the coin that so wishes it was one-sided.

Okay, so I have some terrible news to share with you. Unfortunately, you are not and cannot become immortal. As a result, you will need to face death in your lifetime. The flowers that were used to welcome you once will be used to bid you goodbye. And no, it's not that your body will die. Your body does not die. Your mind with its consciousness dies, and nothing about you can function without your mind and its consciousness. While you and I can logically accept this fact, feelings and emotions always overpower such moments. At the onset of death, it is hard to use reason and logic. We are wired to feel sad at separations, although not all are disappointed. To the wise, death does not come as a disappointment. Likewise, for the troubled. A large number of us lead such troubled lives that we see death as an escape. The sheer tragedy and sorrow of everyday existence are so overpowering that when the moment of separation arrives, it is not that big of a disappointment at all. For the troubled, death is a blessing.

I hope you are not troubled. If for some reason you are, please do ensure you read the last chapter on purpose. I have come to realize that it's not at all worth being troubled. It is an investment with zero returns. I have also found that by sharing what's in my mind, I can ensure that my thoughts and ideas will live on even after I die. Through my writing, you can read my mind and connect with me, long after I am gone. Unless you are willing to share a part of who you are, death will take everything from you. Let it not.

Now then, why are we so caught up with discussing death lately? I have been told that this is the best time to be alive. I've observed that nothing fills the void more than moments of feeling good about yourself and everything around you. How do you get to such states? With money? No, money cannot buy them. One way to feel satisfied and happy is through wisdom. Yes. Wisdom is obtained when you have accumulated the wealth of right knowledge. When your concepts about life are clear, you are on the right road to pursue truth. The most precious asset one can have is a rational thinking brain. The English cosmologist Stephen Hawking once said, "Although I cannot move and I have to speak through a computer, in my mind I am free." I feel that those who have richness through health & wealth can travel to any part of the planet. But only those with richness through high-quality thoughts and great intellect can travel to any part of the universe.

In the words of the famous teacher Frank McCourt, "You might be poor, your shoes might be broken, but your mind is a palace." The intellect gained and freedom allowed to question everything is a gift. Those who open this gift and use it start their journey towards accumulating wisdom. With the power of rich practical thinking, you will see your demons for who they are. Google gets about 8 billion hits a day. Searching for answers is ingrained in all of us. However, like everything else, there are obstacles. Over time, we have lost the ability to ask the right questions. This is because we have created myths and started believing that those are true. That's because our minds are not perfect. As our brains have not been designed by an intelligent creator who could test for bugs before signing off for production, they are not perfect. They are filled with sinister thoughts from time to time. We think about so many things, from good to bad to evil, in every moment of every day. Our bodies evolved to expel all toxins as waste but our minds, unfortunately, did not.

Albeit depressing, I won't be doing justice to this chapter if I don't discuss states like coma. Coma is a state where the mind is alive but apparently no longer responsive or conscious. There is a complete absence of awareness that is observed. Why does the body do this? Well, evolution may have designed it this way, so as to be able to pause things like consciousness and brain activity for a while till the serious injuries heal. So that energy can be utilized efficiently. But, what happens to the subconscious and the unconscious layers? Neurologists seem to talk in a way as if they have mastered this topic and use scales like the Glasgow Coma scale to make everything sound objective. However, I am not convinced. Enough with this topic already. We are hale and hearty.

Buddhism says that the secret to a happy life is to just live in the moment. Do not regret or rejoice the past, nor worry or look forward to the future. Just be in the moment. Obviously, easier said than done. Can one truly be in the moment? How does one define a moment? It is impossible to truly live in the moment, as the moment is too short for you to experience all the countless things happening in and around you. I guess you're happiest when you don't know why you're even alive. Plus, we all need some chaos in our lives, don't we? There is no method to the chaos. The chaos is itself the method. Consciously or subconsciously, we all strive for chaos in life. No one is content living happily ever after. There is boredom in that. We need periods of chaos from time to time. Does this stem from the very innate chaotic nature of the cosmos we were born in? How is it possible for us to have picked that up? Moreover, why did the Greeks choose to name the universe Cosmos, which is in one way the opposite of chaos?

Man: Will you please stop making a mess? Tell me little one, what have you understood about existence?

Wise child: We all are cars that run on fuel, with the fuel being our relationships, hopes, wins, everyday pleasures and so on.

Man: Interesting. What does the car symbolize?

Wise child: The car running is equivalent to your life going on with a purpose.

Man: And where does the mind fit in?

Wise child: Your mind or your consciousness can be considered as the driver of this car.

Man: But, what happens if there is no fuel? The car would stop running.

Wise child: Yes, it would.

Man: What about the driver?

Wise child: The driver would be helpless, as the engine won't work without any fuel.

Man: Where are you going with this?

Wise child: People who have seen and understood the bigger picture of existence are equivalent to cars with advanced engines that do not require fuel to run. They can live in complete isolation, with utmost focus and purpose.

Man: Wow. That's childish yet a powerful analogy.

Wise child: I am a child.

Man: The child takes everything seriously, but the world does not take the child seriously.

Wise child: I see that you've grown up.

While growing up, I had come across the concept of the language of thought hypothesis (LOTH), wherein thought has a language-like structure of its own. Here's a thought experiment: If a man lived alone all his life and learnt no language, what language will his mind think in? Are our minds bound by the limits of the language they are exposed to? Perhaps language is not evolved enough to express everything that our minds perceive and because we cannot put it in words, we probably don't know what it is. Is there a word for this crazy concept in the dictionary? Or do our minds create a language structure and syntax of their own, on the basis of their interpretation of reality? A monk who is silent and has meditated all his life can probably tell us, but he won't talk.

Some monks see the universe as a being forever in meditation. When exactly and why did we start the practice of meditation? The earliest records can be found in the Hindu Vedas dated around 1500 BC. Why did it start? Well, our minds seem to have been cluttered and needing some focus and solitude since antiquity. Why are our minds so not at peace and needing interventions like meditation? In such a finely tuned body with every organ doing its job precisely so that we can survive, why do our minds wander to entertain thoughts that are detrimental to our survival? I don't know and all I can say is that they are connected to the chaotic cosmos and thus mirror its properties. At the same time, I figure evolution wouldn't have allowed organisms to enter such meditative states as it would have made us vulnerable to passing non-meditating predators. So, I don't think dinosaurs entered such contemplative states. But once some survival security was assured, it started. In an ancient era devoid of books and the Internet, it was probably the closest way of searching for knowledge and truth. By figuring it out yourself with absolute focus, in complete silence and solitude.

In solitude, we have some great like-minded company, as solitude is the natural habitat of great thinkers. In complete solitude, we discover our incomplete understanding and try to make it complete. Meditation is but one form of solitude. The mind needs some peace and quiet to be able to make correct sense of the place it finds itself in. So that it is not biased by wrong notions and not swayed away by poorer minds. Rich contemplation in absolute solitude can cause a radical overhaul of our thought process. Let's attempt to exist in the solitude driven sphere of intellectual richness, in order to shield ourselves from the mediocrity all around.

In the backdrop of solitude, curiosity is the mind's greatest companion. It seems to be embedded in us, via our collective unconscious or maybe our collective upbringing, the need to understand everything in and around us. The things around us that may help us understand things around all of us are the millions of species, but we are unable to communicate with them to understand how they consciously experience reality. I think of trees. I doubt if they think. But if they did, what would they make of the world around them? What would trees living in civilizations think of the people around them? What would trees standing alone in the wilderness make of the heavens? If trees could think and communicate with us, their wisdom would have enlightened us to no end. The trees would have told us that we humans have minds with a lot of quirks. For starters, we are born with a sense of insecurity in the way our minds want to know how we've performed and where we stand as compared to our peers and in society. We do not know the limits of our minds and our bodies, yet we choose to only push our bodies. The human engine needs fuel for both body and mind, yet we only focus on feeding our bodies every day. The American writer Mark Twain wasn't kidding when he said, "When I am king they shall not have bread and shelter only, but also teachings out of books, for a full belly is little worth where the mind is starved."

Interviewer: You have quite an impressive resume. What would you attribute your achievements to?
Candidate: I've achieved some things in my life. Some of those have been on my own, and some through the help of others and circumstances. I attribute this to a good amount of dedication and single-minded focus, which in turn are a result of my learning, experiences and genetic construct.

Interviewer: I have come to understand that you were asked to leave from your last job on account of disciplinary reasons.
Candidate: As it turns out, those character traits I just mentioned apply everywhere, including anger and detachment. The passionate mind has no option but to pursue something like rage with single-minded focus.
Interviewer: That is no justification for an act of rage. You have the ability of self-control, which you decided to not use. The more restless and ambitious you are, the more restraint you must exercise.
Candidate: I know what you are saying. But my mind is wired in a way where peace and panic are the binary states of my existence. I am either in your face, or nowhere to be seen.
Interviewer: You come highly recommended and so we are willing to overlook your last mishap. Remember that keeping your mind balanced and receptive always, even with the surroundings impacting it every moment, is your real achievement. The great seas teach us that if we relax, we can float and survive. If we panic, we drown and die. Makes sense?
Candidate: Sense? Let me be absolutely free to be what I want to be, to think what I want to think, and do what I want to do. Rid me of all accountability and let me flourish. I do not wish to be under the pressure of accomplishing anything.
Interviewer: Ours is a free environment for sure, and we won't pressurize you. But we need to know that your deliverables are on track. I feel that is a fair ask.
Candidate: Our ambitions can only be limited by our doubts.
Interviewer: You do have some weird answers to my questions. I haven't met anyone like you before. I am curious to know as to how would you define yourself?
Candidate: Who am I? How do I define myself? I am just a collection of trillions of cells, organized together in a hard-to-comprehend structure. I wake up pretty much a different

person every day. Some days my strengths are my weaknesses, and vice versa on other equally boring days. I long for something, but I do not know what. I am frustrated. I wish to die, but I know that I have to live to find what I am looking for.
Interviewer: Are you depressed? You sound like a person in depression. Is it because of the last job?
Candidate: Nobody is depressed. Some are merely in more solitude than others.
Interviewer: I see. Have you been sleeping well of late?
Candidate: I crave for dreamless sleep, in order to be able to experience a slice of nothing and a slice of death, each and every day.
Interviewer: Listen, I need to have a word with the gentleman who recommended you for this job. Allow me a couple of days for that.
Candidate: Your mind tries to understand what cannot be understood. All phases of life come with an expiry date, including jobs.
Interviewer: Got it. We will get back to you once this week expires.

Does your undernourished mind need a break to refuel itself in order to cope up with this barrage of repetitive questions I am throwing at you? Or do you want to break because your brain is now full? Sorry, you can't for your mind cannot be full. Not in your lifetime it can't. I have looked up and pulled up some figures for your ready perusal (although I believe these are calculated guesses as we aren't smart enough as yet):

- The human brain contains 100 billion neurons
- Each neuron has about 1,000 connections with other neurons

- Information from one neuron flows to another neuron across a synapse
- 1 synapse can store 4.7 bits of information
- Each neuron can support many memories at a time
- Blah blah (in rapidly blurring vision)
- The human brain can store close to around 2.5 petabytes of memory

1 petabyte (PB) is equal to 1,000 terabytes (TB). So, 2.5 PB is equivalent to 2,500 TB of data. That's around 2,500 laptops with 1 TB of hard drive space each. Should be enough! Unless, of course, you are immortal in which case you will eventually need an upgrade or a stressful cleanup.

I suffer from motion sickness, and that was at its peak when I was young. I hated traveling in cars, especially suffocating diesel engine ones, as I would keep throwing up and feel miserable. I understood it to be a medical condition and ate medicines to control it while traveling. It is only when I grew up and had access to the Internet to research, did I realize that motion sickness is just a survival mechanism that the mind employs. When traveling, the body is stationary inside the car and the car is moving. On account of this, the brain can sense motion but realizes that it is stationary. This leads the brain to enter a state of dilemma and concludes by thinking that it is hallucinating on account of poisoning. So, the body throws up to remove the poison. I may have been naïve not to have known this already. You may have already known of this considering the smarty-pants you are, but I so had to mention it here for I was simply amazed by this when I read this some time back. It's incredible to note the things the mind is equipped with to keep us alive, and so much of it keeps happening subconsciously.

Very well, where were we with the random flow of simple thoughts? We have so many shortcomings. We fail to see the simplicity of everything that surrounds us and search for things we cannot see. We want to see our inner self with mirrors that don't exist. Our minds have a belief system filter that rejects or ridicules information that is inconsistent with our beliefs. We all want to feel special internally, and search for acceptance and recognition externally. The plain vanilla reality is all too boring for us, and we invent the mystical to entertain us. The pattern recognition power in our minds helps us see a helical structure in our DNA, and yet we are more focused on finding the face of God in clouds. In the face of all of these shortcomings, our minds can very well choose to play victim. Funny things those, our minds. They are flawed and yet phenomenal. A mixture of beauty and ugliness, complexity and simplicity, good and evil, elegance and mess. That, ladies and gentlemen, is the mind. Its imperfect construct is only matched by its desire for perfection.

It is the imaginative mind that experiences revelations and direct conversations related to divine providence. Back in 1917 Portugal, some shepherd children had a revelation that the Virgin Mary would appear on October 13th and perform miracles. This sparked a great amount of curiosity and thousands of people gathered to witness this supposed miracle. Once the theatrics were over, some people testified that they indeed saw miracles in the form of bizarre activity of the Sun, wherein it danced and emitted rainbow colors in the sky. It is the human mind that convinced itself that it saw this entire supernatural phenomenon in the sky. The mind can choose to create, see, and believe its versions of reality. Thus, it's not reliable right? But without it, what are we?

Nothing, we are nothing without our minds. Our perception of reality is through our minds. Everything in this book is null and void in the absence of cognition to understand it. It is the mind that sees the glass as half empty or half full. Without the mind, there is no glass. In a metaphysical way, our reality is a creation of our minds. Does a thing which is not in any of the three layers of my consciousness, i.e. a thing I don't know that I don't know about, actually exist or not? For me, its existence may have no meaning, but for others, it may. It is part of their reality. Look at me, talking as if there are multiple realities floating out there. My mind wanders out of control when mulling over such deep subjects.

Is anything in our control? Yes, of course. By getting on the right road and pursuing the absolute truth, we can obtain the gift of wisdom. You can't try filling a cup that is already full. Thus, always leave room in your mind for the right answers. Forever remember that the omnipresent patterns and cycles around us are invisible to the mind that's unwilling to wonder and be humbled. Train your mind to experience nature and reality the way they were meant to be experienced. Don't experience the beauty of the rain by watching raindrops fall on the ground, but do so by becoming the rain instead.

VALUES & SOCIETY

All truth passes through three stages. First, it is ridiculed. Second, it is violently opposed. Third, it is accepted as being self-evident.
– Arthur Schopenhauer,
The World as Will and Representation (1819)

There is something wrong with how mankind has constructed societies. Our design and understanding of values are completely wrong. Morality is a man-made concept. The cosmos has no notion of values, ethics, or good deeds. Comets follow no path of righteousness. Nothing happens by design and karma. Random explosions consume stars and planets. Stray asteroids wipe out civilizations. There is no judgment passed in the cosmos. There is no concept of right and wrong. That makes ethics or morality an artificial concept restricted to beings on planet Earth. Small and insignificant.

Yes, ethics and values are artificial. We have chosen to construct societies based on morals where some actions are considered as sins. I agree that such concepts are required or else we'll all rob and kill each other. Moral law and order are required. But we should be clear on why it is enforced. We seem to have never understood the reason correctly. Through our upbringing, we see good deeds as the right way to live because we want to be good human beings. But, the concept of 'good' is something we made. It is not naturally occurring and thus, the same applies to 'bad'. There is no good or bad. We made all of this up so that we could survive together in societies we built. Yes, survival is important and if values can help us with that, they come into play. At the same time, we should see them for what they are, instead of seeing them as something that's ingrained in the fabric of existence, and in the foundation of our species. They are absolutely not. Please note that there is no ethical need for the universe to exist in the first place. In fact, the universe does not understand the very concept of ethics.

Wise butcher: What would you like today, sir?
Man: You should close down this shop.
Wise butcher: Why do you say that?
Man: You are a killer. My heart goes out to the baby lambs and chickens that are slaughtered for meat every day. They roam around in slaughterhouses playfully, full of curiosity and wonder about the world around them. Waiting for their mothers to play with them, completely oblivious of the fact that they will be killed in a few hours. What right do you have to take a life?
Wise butcher: Many people turn vegetarian because of such thoughts. But remember that the food chain is not based on ethics. Nothing in nature is. Humans have eaten meat for eons and will continue to do so. We are members of the food chain and that is our role to sustain the balance in nature. You should know that good and bad are artificial concepts in this natural world.

The dinosaurs lived on the same land we live on. They did so for 180 million years. I am sure they had an unwritten but understood moral set of codes for them to co-exist and carry on. They formed their groups and had their mental definitions of right and wrong. But then they all died. Did they all commit a collective crime to earn that fate? Certainly not. They just happened to be at the receiving end of a natural occurrence. The thing we can learn from the extinction of dinosaurs is that a random event in the universe can wipe us out. It does not matter whether we do the right things or not. The universe is neither benevolent nor hostile. It is thoroughly indifferent.

When there were no groups or societies at all and it was just a game of survival of the fittest, values and ethics had no meaning. It was all down to who was able to find food and not get eaten. Nothing else mattered. When some amount of structure came in and societies were constructed, there automatically arose a need for moral beliefs to come in so that some safety and order could be ensured. Again, I am not saying that the concepts of good and bad don't matter. They matter in society as they help us not do barbaric acts. But, the understanding of why we follow those concepts also matters. We should always know, at the back of our minds, that it is to keep the house in order. The house is not made of values. It is made of indifferent bricks.

If a mother bear gives birth to deformed cubs, she will most likely kill and eat them. That's correct, a mother eats her children. The mother bear does not fear the value police of nature or society punishing her for this sin. For her, it is common sense. She sees it correctly and so does a tiger. A tiger does not care for the innocence and hope in the little deer's eyes. It kills it and eats it, as the tiger needs to survive. Carnivorous plants like the Venus flytrap also have figured this out. When nutrients in the soil are not to be found, they trap and consume unsuspecting inspects. The meanings of virtue have no meaning. The animals and plants in nature seem to have got it right. While we humans have evolved intelligent brains to get it all wrong.

Back in 300 BC, Alexander the Great created the largest empire of the ancient world, stretching from Greece to India. While growing up, Alexander was educated by the great philosopher Aristotle who taught him morals, among other things like logic and religion. Alexander was greatly inspired by writings on ethics by Aristotle. What then led him to lead an army of men that killed close to a million people in their quests? Civilizations were burnt, people were enslaved, and happy families were destroyed. I am sure there must have been countless innocent women and children in those massacres. How did such an immoral act go unpunished? Forget punishing, our value-driven society started calling him a hero. He is considered one of the most important people in history and military organizations worldwide still teach his tactics. He married thrice and led a fulfilling life full of accomplishments and rewards. What am I supposed to make of this? That either killing the innocent is outside of the value system and is acceptable in the human society, or that we selectively accept that we are like animals and it is indeed survival of the fittest. There are double standards for sure.

I have been so worked up about this that I can't help but do some extensive but disturbing research and pull up more such worshipped figures from history who have allegedly done immoral deeds. Note that by doing so I don't intend to provoke or hurt the sentiments of the followers of these icons, but rather open eyes of people who are willing.

- Leonardo da Vinci: Had intimate sexual relations with his pupils and engaged in sodomy.
- Thomas Edison: Made inventors work in his workshop (including Nikola Tesla) and called their inventions as his own. This included the light bulb.

- Akbar the Great: Carried out a merciless massacre of 30,000 innocent civilians; created a tower out of the heads of victims on another occasion. He had a total of 300 wives and concubines for his pleasure.
- Sathya Sai Baba: Many have testified that the guru sexually molested many young boys in the name of healing and correcting energy.
- Christopher Columbus: Enslaved innocent and friendly people of the Taino tribe in West Indies and subjected them to extreme brutality.
- Oscar Wilde: Solicited 12 boys to commit sodomy.
- Tipu Sultan: Destroyed local religious places of worship. Forcibly converted the religion of thousands of people. Killed tens of thousands of people in the most barbaric manner.
- Charlie Chaplin: Was a known child offender. Impregnated two girls who were under the age of 16, and married them so that he could escape being prosecuted.
- Mother Teresa: Carried out forced conversions by baptizing suffering Hindus and Muslims to Christianity without their consent.
- Osho Rajneesh: Amassed a huge amount of wealth from his followers, which he used for his personal welfare—he owned more than 90 Rolls Royces. His practices included carrying out drug-induced orgies in his ashrams.
- Mahatma Gandhi: At the pretext of testing his chastity, a lot of indecent experiments were carried out in his ashrams across western India. I am not going to get into the specifics here. You can look up the Internet for excerpts from a book on him by Jad Adams.

The last one is most shocking, once you get into the nitty-gritty. I encourage you to do your own research by looking up all of these cases on the Internet. As most of these cases are not explicitly proven and never will be, for whatever reasons, then why am I speaking ill of the dead? Well, all of these icons (and many more I don't know about) are spoken of with high regard in the moral circles of society, and none are remembered for these controversies or disgraced for the evil deeds they most likely committed, or the lifelong trauma they must have caused to those affected. Why so? I am trying to figure out how our value-driven society works.

The Germany Nazi regime killed 17 million human beings during the time of the Second World War. This included 6 million Jews, the elderly, sick people, homosexuals, and the disabled. The leader of the Nazi party, Adolf Hitler, considered all of them as part of an inferior race and unfit for modern society. People were taken to extermination camps and killed in gas chambers. All kinds of horrific medical experiments were conducted. Twins were killed just so that their bodies could be dissected and studied. I won't lie to you. I stopped researching this topic midway, as I just couldn't continue reading the kind of sickening atrocities that were carried out, especially to innocent children. I cannot even imagine the suffering those 17 million people must have gone through. My mind drifts away to the image of a parent begging to an evil leader to spare the life of his child, only to then witness the child being mercilessly slaughtered. Those were ghastly times.

Hitler enforced a survival of the fittest regime himself by passing judgment and killing whoever he felt was unfit. For all that he did, he lived a happy life once he became a political leader. He loved and married, and also carried out an illicit relationship with his half-niece. He enjoyed the admiration of his men and the satisfaction of having achieved success in life. He lived till the age of 56 and when in danger of being arrested and tortured, he took the easy route of committing suicide.

Where is the justice? In my eyes, he committed 17 million heinous crimes and walked away scot-free with a smile on his face. The moral society let him.

While I am angered and upset about how people like Hitler can get away, I have to put my sane reasoning hat on and try to understand why Hitler was this way? How do human beings, who are born in God-fearing and value-driven societies, turn out this way? In the previous chapter, we talked about psychopaths and how their minds work. But Hitler wasn't a psychopath I think. He was a sociopath i.e. a psychopath not by birth but by upbringing and circumstances. He was born in a poor and negligent household where three of his six siblings died in infancy. He was frequently beaten up by his father, and he constantly fought with him. He held a grudge against his father and did poorly at school intentionally in order to get back at him. He turned into a detached and sulky child. He did not get admissions in the schools he wanted to and by the age of 20, was homeless and worked as a casual laborer. This traumatic phase must have sown the sociopathic seeds in his mind and defined the set of values he was to follow all his life. I can only imagine the anger he must have developed and repressed. Like all sociopaths, he must have become fearless, acquired good acting skills, developed a high tolerance for stress, and become incapable of feeling any emotions or regret. These traits helped him become a super-achieving political leader. Once in power, his sociopathic slaughterer instincts and repressed anger came out in full force. Did he see those 17 million murders as completely moral acts and in line with his values? Perhaps society failed by electing a delusional sociopath as a leader. In such matters, maybe it is good if royal families always take charge and each well-brought-up family generation is in command, instead of picking leaders from the local masses with possible troubled backgrounds. In the corporate sector, psychopaths are bound to make good CEOs but there thankfully are policies in place to keep any malpractices in check. I know of a maverick CEO who is a psychopath for sure. My mind wanders.

Sticking to the subject of values, here's one more example from the past. Slavery was like an established institution back in the day. Would it be unethical to compare it with poultry farming of today? I guess not. Nobody saw bad in it for it served a purpose in the economy. Oh, but we are not talking about chickens. We are talking about our fellow brothers and sisters. The greatest sin our moral society ever committed has to be slavery. Prevalent across the world, and belittling the existence of millions, this practice made sure that inequality and abuse would forever be embedded in the roots of our societal foundation.

How can humans carry out this kind of abuse? Non-performing slaves were punished by burning, mutilation, whipping, hanging, and rape. Sometimes, this was done just to assert dominance. Infants born to slaves were used as baits in the rivers to lure alligators. Oh, the sheer horror of it. What value-based society are we talking about when its members are capable of such insanely cruel acts? Plus, where is the accountability and what about fairness? Such brutal slave owners led rich happy lives and died peacefully. Who is paying for their sins? Why should I lead a tough moral life if there is such injustice going on elsewhere? Phew. This is getting to me now. Breathe. You know, the human mind is to be blamed sometimes for such continued acts of torture. Torture, for a person being tortured, becomes tolerable and a way of life after a period of time. The mind adapts to think that all is well so that the person can survive. So, the tortured person never stands up to revolt and makes peace with it, thereby allowing the torturer to get on with his merry ways.

A wise man once shared his thoughts with me on why injustice and inequality are actually sometimes beneficial. It is based on the fact that we humans do not know our abilities until we are stretched and pushed to our limits. When certain humans are subjected to hardship and oppression, they break all shackles and become super-achievers. Values like fortitude and dedication come to the forefront and enable these individuals to fight the odds and achieve extraordinary feats. If not for the hardships, those abilities would have laid dormant and these individuals would not have gone on to become role models for others who would then go on to display these values in their own endeavors. Thus, humanity as a whole progresses with such inspirational stories. So, one can argue that an ideal society would not have any super-achievers and thus won't progress. It's a double-edged sword.

We were discussing how the notion of values may have edged its way into the animal kingdom. Although like I had said, the tiger doesn't give two hoots about morals when killing a baby deer for food, it does seem to follow some values when living with a group of tigers. Notably, by being nice to them and not hurting them. Why is that the case? Why this selective use of goodness? Evolution tells us that this is on account of selfishness. You heard me right. If there is one trait that has helped organisms survive and evolve, it has to be selfishness. We can also call it self-interest but I am going to stick with selfishness. A tiger knows that there is safety in numbers and better chances of finding food. Thus, ensuring its survival and chances of reproduction. So, it allows other tigers to join and share resources in a quid pro quo arrangement. As a result, they don't fight with each other for supremacy or resources. So, the visible value called teamwork or empathy or friendship is actually pure selfishness to ensure survival. The same logic holds good for birds sitting on cows, or specific symbiotic relationships like the egret bird and the buffalo. They are not doing that to promote world peace. They both are in it for their selfish interests. You scratch my back, and I will scratch yours. I will be nice to you, so you will be nice to me—we'll both help each other out and ensure we both survive. So, we can so clearly see that this entire opera of evolution's survival of the fittest theme is just selfishness of various creatures acting in cohesion. It's like an intelligent riot.

Now that I think of it, many of the meaningful accomplishments of the human civilization have also been a result of selfishness and material desires. From the Americans landing on the Moon to show supremacy over the Russians, to inventions happening for recognition, competition, and rewards, we owe the advancement of the human civilization to selfishness. In fact, America canceled all planned manned missions to the Moon and beyond once supremacy over Russia was established. It is not as if we attained nirvana after landing on the Moon and haven't strived to venture any further ever since. It is purely because our selfish interests were met. Knowing this, it is natural to question why our society sees a selfish act as a bad deed, and it is frowned upon. We are discouraged to indulge in it. But we are a product of evolution and have made it so far because of the selfishness of all our ancestors. So, we are bound to be that way. My wanting you to appreciate this insight of mine is my selfishness on display. While I have convinced myself not to think this way, but I am only human. Plus my moral maturity varies from time to time.

Also, although a controversial topic, we can choose to see forced sex or rape in the same light. How can men even think of such awful acts? I'll tell you how. It's because evolution made their brains to think that way. I am not taking their side, oh please no. I am just trying to figure out why such immoral thoughts can even be conceived. Our diverse animal kingdom has creatures like ducks, geese, and chimpanzees, which have been observed to frequently engage in forced copulations. Yes, it's proven. Why do they do it? Well, for supremacy and for carrying on the bloodline. Now, we are not imprisoning those chimps, are we? As we have evolved from the same set of creatures and are all connected, we should know where those horrific tendencies come from. Nature and evolution gave it to humans.

And that's why I think the concept of God was brought in to dissuade humans from entertaining such thoughts or doing bad deeds, else they would pay for their sins by burning in hell. That leads us to an interesting question: Is the fear of God necessary for morality? That is, in the absence of God and moral teachings of religion, will we be an immoral society? Well, evolution did not work in tandem with God to instill the values that we just discussed. I seem to be opening a Pandora's box here, so let's defer the miracle of (and called) God for the next chapter. For the time being, we'll look at one example of how the fear of God's wrath has been used to instill morality in society: In the Hindu Sanskrit epic called the Ramayana, the legendary and ideal prince Rama shoots and kills the monkey king Vali, who is the elder brother of vanara Sugriv. A dying and somewhat perplexed Vali asks Rama:

dharm hetu avatarehu gosai.
marehu mohi byadh ki nai.
main bairi sugriv piara.
avagun kavan nath mohi mara.

Translation:

Though my lord, you are an incarnation of Dharma,
You have shot me like a cruel huntsman,
Am I your foe, and Sugriv your dear friend?
For what fault, my lord, have you taken my life?

Rama makes Vali remember all the sins he committed in his lifetime, which include never following any moral values, mistreating his brother Sugriv by making him homeless and robbing his wife Ruma immorally, and leading a lowly and treacherous life. He also preaches the following:

anuj badhoo bhagini sut nari.
sunu sath kanya sam e chari.
inhahi kudrishti bilokai joi.
tahi badhen kachhu paap na hoi.

Translation:

A younger brother's wife, a sister or a son's bride,
Are as one's own daughter. Listen, O wretch,
Whoever looks upon these with a lustful eye,
May be slain, without any sin.

Rama says that he is justified in killing Vali as punishment for his sins and lack of adherence to values. The Hindu religion and society are founded on the stories and messages of such epics. As one would expect, the needle of the moral compass must forever be in the right direction for people living in a society founded on such stories.

Like everything else, values also evolve with time. Not through natural evolution though, but rather enforced by man. This means that people's perception of some of the values changes with time. Hunting of animals was an activity of pride and satisfaction in the last century. Today, it is a crime and brings shame and regret. Our pious ancestors saw no fault in constructing obscene sculptures in public monuments. Now, any such public display is frowned upon. With time, our levels of shame seem to have increased and the moral value of decency catapulted ahead. As social and economic circumstances change, we tend to slightly modify the definitions of good and bad so that they help us survive and prosper. Humanity is supposed to benefit from having them. This again goes to show the artificial nature of values. If they were naturally occurring like physical laws, we wouldn't have been able to change them. But they are not. So, let's stop treating them like some universal commands that need to be obeyed in order to live a good and fulfilling life.

I think about how intelligent civilizations living on other planets would treat something like values. A species that is more intelligent and advanced than us will have nothing like a moral value system for sure. Those beings would have no notion of good and bad. They would do things that make sense and those would be the right things for it has gotten them this far in advancement and intelligence. I imagine that they would be kind to each other and work collaboratively without prejudice, and they would do this without thinking of any of these as societal requirements. They would be absolutely indifferent and full of goodness at the same time. And, they would do all this without the fear of what non-conformance would bring upon them. I wish we go on to become like them one day.

I reiterate that I have no doubt in my mind whatsoever that our cosmos has no notion of values, ethics or good deeds. We have created these artificial notions in societies we have built, and we are now ruled and defined by these values. At the same time, we are thoroughly capricious in our approach, as we have seen in the unspeakable examples that I have listed earlier. In a way, we have created what we believe are civilized societies but the members of these societies are still completely filled with primitive wild instincts, as provided by nature and evolution. So, we are trying to pretend to look civilized and fooling our minds into believing that we actually are something we are not. The tiger is trying to make its mind believe that it is vegetarian. Where does that leave us? Morality is indeed a man-made concept and in line with the imperfect and primitive beings we are, we do not know how and when to follow it. Also, we either do not know or often keep forgetting the real reason why we have invented morality in the first place.

Man: How are you?
Wise slave: You seem to be in a kind mood today. Did your God pay you a visit?
Man: Brother, I am very sorry. I am to be blamed. I have realized my mistake. I have no right to mistreat you and keep you as my slave. I am overcome by regret. I wish to kill myself. My western civilizations and I have robbed the world for long. We think of ourselves as a superior race and look down upon other races. I seem to have forgotten that my ancestors arose from the African savanna.
Wise slave: We are all creatures made by nature and evolution. The game of survival of the fittest plays in nature all the time. If you won't fight, you will be killed. I have ended up a slave because I fought and lost. I cannot blame western civilizations for my misery. It is for me to rise and fight to win my place. I cannot sit here sulking and expect nature to come rescue me. I need to find a way to survive. My children need to be born and they need to live.

Children can help solve some of the puzzles daunting us on the theme of values. The best way to know the list of values we are born within our innate evolution-constructed-natural-self is to study the values that children display before they are taught about good and bad. Experts in this field have conducted interesting experiments with babies and toddlers to understand how they behave and react. As it turns out, they observed that babies and toddlers displayed tendencies to favor the good guys and reject the bad guys. In other terms, they displayed small amounts of moral sense and fairness. So, this trait of being nice to people who are, or will be, nice to us seems to be ingrained in us right from the time of birth. This is on account of possibly a collective unconscious that we all share, a topic that we had discussed earlier. As it turns out, being nice because of selfish reasons is a trait that was embedded in us by evolution, and all the rest is injected in us by society.

Once we become aware as a child while growing up, we find that there's something already inside us although we don't know how. It's called our conscience (not to be confused with the terms conscious and science). Our conscience is the internal compass that tells us right from wrong, and how to go about adhering to values. It's what makes us feel guilty. Where did this nagging conscience come from? In addition to being a moral compass, I also like to think of conscience as an internal filter through which all thoughts need to pass before they become actions. So, if I get an evil thought, my conscience will reject it and I won't translate that thought into action. Likewise, if the thought is good, then my conscience will happily stamp and approve it. How did evolution manage to pull this off?

It turns out that conscience happens to be related to our sense of morality. It's because we developed the value of being nice to other creatures in our group, our minds needed to know what action was nice and what was not. Thus, a primitive version of this self-control filter was put in place by evolution. With time, we humans have reinforced and solidified it through social and religious teachings. The guilty conscience has grown stronger with time. My conscience pokes and tells me that I have done a horrible job at explaining the origin of conscience. Noted. This deep mystery stumped even the likes of Darwin and is still being figured out. Pardon me.

A wise man once told me about how he was made to join a group that was singing and chanting together. Although he would not have done that had he been alone, but he felt a strong sense of pull and drive to sing and chant when he was in the group. In a way, being in a group changed his thought process to an extent. Do we think and act differently when we are alone as compared to when we are in a group? I know for a fact that runners perform differently when running alone as compared to when running in a group. We find jokes funnier when sitting in an audience as compared to listening to them alone. The group energies definitely rub off on the individuals. Also, are our values and definition of right/wrong guided by what the group collectively believes in? The experts say that yes, it does. The group develops a collective mind of its own and all the individual minds align with that. We have seen examples in nature like ant colonies and beehives, where all organisms live their whole lives together in the group. The group displays a collective set of traits, values, and beliefs, and all the members of the group mirror those. The obvious question is that why do individuals act this way?

We have already seen how and why we chose to live in groups. During the times of our distant ancestors, like-minded folks in groups ought to have formed a majority, and when there was a difference of opinion, the majority must have prevailed. Those who opposed the majority would have been killed. So, there was a survival advantage if an organism started thinking the way the majority of the group members did. As every organism in the group learnt this fact, the whole group became one organism in terms of having a single collective mind displaying one set of values and beliefs. We can see this playing out in societies and religious communities as well. The collective set of values takes on the form of the acceptable list of values that need to be taught and followed. Anything that's not in accordance is considered immoral. The interesting thing to note is that none of this requires a leader planning and shaping any of this. Such things take shape on their own.

I don't think there is any institution where values are considered more important than the armed forces or the military. After all, we are asking soldiers to not think twice before dying for the sake of the country. That needs some serious belief by the soldiers in values like selflessness and loyalty. Their stories are used as case studies and examples to drill down values in other walks of life. It is another matter that I completely disagree with the concept, but we are going to talk about that a little later. For now, we can see how values play a pivotal role at the individual level in meeting the objectives of the larger group. Also, I do not encourage you to join the army.

In the corporate world, organization leadership values the need for corporate values in order to build world-class organizations. These are typically a set of around four to six workplace related values, which all employees are expected to demonstrate. They are along the lines of integrity, commitment, teamwork, and the like. The valued function of human resources works with the executive leadership to arrive at the set of corporate values, in line with the corporate objectives and the desired corporate culture. Why am I boring you with these particulars? Well, in order to inform you that this is a corporate exercise in futility. Values cannot be instilled in employees by training them on those, frequent awareness efforts and posters on the walls. First and foremost, employees are people and people walk in with their own set of ingrained values on account of the lives they have lived and their upbringing. They will demonstrate those values, and if some of them are not in sync with the corporate values, then that's that. Second, corporate values are built on their own by how business and work are carried out. The workplace culture is forever built in the background and takes form on the basis of how managers treat team members and empathize when needed, how much importance work-life balance is given, and what the management priorities are. It all depends on whether all of the focus is on meeting the end goal revenue targets, or if the focus is on how work is carried out and whether processes are adhered to and ethics are given importance. You get the picture. If you are part of corporate leadership, now or in the future, please keep this in mind.

How values take natural shape in the corporate sector turns out to be an example of how things worked in various civilizations as well, and in turn in societies and countries. If the political leadership has the right priorities and objectives, then the intended culture and values are formed automatically in the background over time. They cannot be built separately. It's all connected. It makes sense to quote the great economist and philosopher, Adam Smith, in this context: "Little else is requisite to carry a state to the highest degree of opulence from the lowest barbarism, but peace, easy taxes, and a tolerable administration of justice; all the rest being brought about by the natural course of things." Thus, great economies with happy people living value-rich lives take form naturally, provided the climate is right.

As is evident, we have digressed from trying to understand why we follow values to the activity of preaching how societies should build values. My bad. Our minds seem to be getting too engrossed with values and morals. Exactly like how our societies are more worried about how things are being done, rather than what is being done and why it is being done. Yes, you scratch my back, and I scratch yours. But shouldn't we, one of these evenings, sit down to figure out why our backs are always so itchy in the first place? We are on the lookout for the truth about the cosmos, and this truth is beyond the concept of morality, for it does not adjust itself to fit with societal norms or beliefs. It is way greater than anything, and also those who will hopefully unearth it one day.

As we continue on the road to finding this elusive truth, I will keep you company. I have some more thoughts to share on society that should interest you, I hope. I have a lot of problems with the kind of societies we have built, our beliefs and practices, our thought processes, and everything else. The sum total effect of all human existence so far is that we are all conditioned to think and act in particular ways, which I think is utterly wrong. I know, I complain of the traffic while I sit in it and am part of the problem, but I think many don't even know that there is a problem. So, it is my job to share and lure your careful attention. Thank you kindly.

All that glitters is most of the times gold. Gold is an object of desire for many, and out of reach for most given its cost. You should know that it has no medicinal or any other survival related benefits. Yet, we long for it. That is because it is limited and thus exclusive. People like to wear it and flaunt it. The total amount of gold on this planet (including mined and yet to be mined) is around 250,000 tonnes. I have done some rudimentary math to calculate how much gold each family on this planet can have and cherish. We can assume that there are around 2.5 billion families on this planet, which is a good estimate. That means that each family can possess about 100 grams of this rare metal that was formed in celestial stars that died billions of years ago. But that ain't going to happen. Prices are kept high so that not everyone can afford it. Plus there are no limits on how much of it you can buy if you have the money. Our mental construct, through social conditioning, is that we should have things that others don't and we should make it a point to show everyone that we have them. The longing for exclusivity by human classes is the bane of our civilization. As a sign of protest, I refuse to wear any gold. It's a different matter that I don't have any.

The treatment of gold by our dark society sheds light on the practice of having restricted access to maintain exclusivity. From people of certain races not being allowed into places like clubs and libraries, to people of lower castes not being allowed into temples, some humans seem to have this rotting weed in their minds that they are more superior and pure than others. In their minds some things cannot be compromised, come what may, and so all inferior and impure must be kept at bay. Seriously, what kind of a society have we built? I feel ashamed to be a part of this farce. The only solace I see is that these 'lower' races and castes will eventually revolt. The inevitable result of allowing restricted access to libraries is that when the angry excluded group turns up with fire and weapons, there are few people to protect those libraries. Thus, justice is supposedly done. But, in the end, books are burnt and progress is undone. Who is responsible for this?

Why do we subconsciously derive pleasure in seeing others fall? Others getting robbed or their cars getting damaged in accidents are a great source of amusement for us. We don't consciously do it, but it seems to be subconsciously ingrained in us. Why is that so? In a moral society built on supposed goodness and empathy, why would our basic construct be so flawed? I think that this is on account of our evolutionary selfish instincts coupled with modern day social conditioning. The marriage of these two leads to a mediocre mindset, which is something I like to define as an intellectually poor mind thoroughly devoid of wisdom. Yes, mediocrity is all around us. A lot of people look intellectual and wise, but the mediocrity comes sailing through via their habits and when they start talking. They neither know that they're that way nor choose to be that way. It's just that that's how people and conditions around them have been all their lives, so they are permanently conditioned that way. This includes the wealthy as mentality and thereby habits are not governed by how rich one is. People want to feel good about themselves, feel that they are valued and their life is on track. They lack the bare minimum wisdom and are occupied with petty matters including showing off and judging other people.

Man: I have no money to give you. I am repaying a housing loan, saving up for an international vacation my family wants to go on, and trying to save up enough to feed my family in case I get fired from my job. Please ask someone else.
Wise beggar: People find comfort and pride in knowing that they have arrived in life. That they have bought a house, have solid investments, a happy family and travel to exotic locations. They feel pride in showing off their happiness, and hoping that others get jealous. It's quite sad, don't you think?
Man: I know that and you are absolutely right. But I am completely immersed in and overpowered by the societal definitions of achievements and happiness.
Wise beggar: The problem is that you seek identity, which means some connection with the society you live in. You seek acceptance and appreciation that you are a successful contributing member. All of this is why you live in misery.
Man: What can I do? Please guide me.
Wise beggar: It's a choice you have to make. My advice would be to strive for intellectual wealth and the resulting happiness.
Man: Perhaps once I have saved enough, I can do that?
Wise beggar: No amount of wealth can buy the intellectual richness quality thoughts bring. The more you delay it, the more conditioned you will become.

Man: I am helpless for the time being. Success in our society is currently measured through material wealth and not intellectual wealth.
Wise beggar: As that measure itself is wrong, you should have nothing to do with people taking such measurements.

Striving for wisdom is difficult no doubt. You become the odd man out in a society working in a particular materialistic way. Your parents are conditioned, and so they expect you to have achievements that they can talk and show off about. It's difficult, I know. You can't stay in a rented house all your life, can you? Your wife and kids cannot always be under house arrest, so you need to budget for holidays. Plus, there is inflation eating into your reserves. Yes, it's difficult, but not impossible I would imagine. If it can give me wisdom, I will go for that road and figure a way of dealing with the obstacles. I will bank on my intelligence and ability, both gained through eons of evolution, to help me out. I would think it would be the expected price I would be paying to get out of the societal rut and the sorry artificial nature of my existence. I like to equate the mediocrity of everyday matters with the force of gravity. It pulls you downwards. For wisdom to be obtained, the escape velocity must be reached and maintained. Ready for launch?

One of the most hard-working people in the world would have to be the Japanese. Their work ethics and standards are second to none. But their societal norms and peer pressure are also unparalleled. There is tremendous pressure for perfection in all matters, right from saying hello, making tea to error-free business reports. That society has norms around how hard one needs to work, the values that need to be followed and the etiquette that needs to be exhibited at all times without fail. There is no room for error. You make a mistake if you are not perfect. If people get fed up with it and stop conforming, they are shunned out. There is only one road and you need to be on it. That must be a very difficult society to live in. Something needs to be done I think.

Judging others is space-time wasted. In comparing and judging, the fools inside us have their say. Even though that's the case, we are at it all the time. Why do we do it? We have seen that the deep-rooted competitive survival instinct in us makes us want to be better than our peers. But why don't we leave it at that? That is, run the race and move on. Why are we forever obsessed with what others are up to and figuring out why they act the way they do? Also, why do we derive pleasure when others fail or perform poorer than we have? Did evolution give us this horrible habit? Yes, it must have. Feeling better about ourselves helps us become confident and increases our chances of survival. So, if we can feel better by thinking of others as lesser, then we will always do that. Anything for survival, you see. But surely, there have to be better ways. Passing judgment on people is so petty and cheap. We are the result of billions of years of laborious evolution. We can do so much better. If we can't, then we just need to make this feeling extinct by completely ignoring what we hear. That's what I try to do. I am indifferent to what people who know me think of me. Discussing people and judging others is the lowest quality use of time and energy, neither of which we have much to spare. Also, I request you not to judge this book or my amateur writing. Read for pleasure and to gain a better understanding of the reality you live in. That being said, don't you forget to leave a positive review! Just kidding.

Man: Can you show me something in the 5,000 range?
Wise jeweler: Is it a gift for someone?
Man: Yes, a wedding gift for a friend. He gave me a gold bracelet on my wedding.

Wise jeweler: So you are obligated to gift him something equally valuable right?
Man: Yes. That's the society we live in and those are the norms.
Wise jeweler: Not really. If you don't agree, you shouldn't follow them. At the end of the day, it's we who make the society and write the norms.
Man: How exactly do I break free from the social bonds and norms that I don't agree with?
Wise jeweler: By standing your ground and choosing not to comply.
Man: But that will not only affect my relationship with my friend, but also risk making me a social outcast.
Wise jeweler: If your friend is wise, he doesn't care if you give him a gift or not. Plus, you should be okay with risking being made an outcast if it can help you obtain wisdom and get you closer to the absolute truth of the grand cosmos. That's all that should matter to you.
Man: I have been a part of this society since birth and norms are so entrenched in me that it is impossible to just cut off.
Wise jeweler: You are not cutting off and entering exile. You are intelligently and selectively not complying so that you can focus on more important things. You will continue to coexist in society but will also guide it forward in the right direction. I know that people have given you things over your lifetime, which you did not ask for. As a courtesy of societal convention that they follow, you did not refuse those gifts. But those gifts, in no way, make you obliged to proactively keep in touch with those people, and that is perfectly all right. The wise do not keep track of such things, and the fools are not worth being in touch with.
Man: Your words are solid gold, my friend. We live in a materialistic society filled with delusions and ego. We have made it that way. It will change when we change.

A pet peeve of mine is people making small talk. 80% of the world's population seems to have a problem with silence. There is an uncontrollable urge to talk about something and stop the restlessness that silence seems to cause in many people. Thus the mouth opens and something random read about in the newspaper that morning comes out. The net result is small talk that's not useful for anyone and kills the rich environment in the making. We never seem to have realized the kind of peace and comfort silence can bring. Silence is the perfect environment for rich thoughts to be entertained. Why do we have such a problem with silence? Is it because people just absolutely cannot exist in solitude with just their thoughts for company? I think this is a symptom of a bigger problem. We want to be forever occupied, and hence a majority of us choose to accept problems in our lives so that we remain occupied and can stay away from coming face to face with our thoughts in a state of isolation. Being left alone with only our thoughts for company is too daunting a task for most. Let's mull over that in silence for a while, and when faced with the urge to make small talk, let's choose to remain silent instead. The world will be a better place.

The training arm of the corporate human resources function often employs icebreakers to kickstart training programs. Trainers, most of whom are born with sunny dispositions, use these games and activities to help ease some of the initial social stiffness and to get participants acquainted with each other so that a collaborative learning environment can be established. While that activity serves its purpose, I have many a time wondered why we are conditioned to feel like strangers unless we interact and why is there social awkwardness that starts building up in the first place? Why do we feel that there is a need for external intervention to help connect with each other? As we have learnt, we are all related at a fundamental level and share so many things in common, including our lineage. We should feel completely connected without any kind of interaction. Also, such icebreakers are not activities that introverts look forward to.

Society has somehow managed to bifurcate and classify things as complex as the mind and personality into types called as introverts and extroverts. Introverts find comfort and energy in solitude while extroverts recharge their batteries by being with people. Why would evolution wire the brains differently to create two personality types? Isn't introvert the default choice as being silent helps find prey and escape predators? Perhaps being extroverted helped organisms to be more popular with the opposite gender and thereby increased chances of reproduction. But they were also vulnerable, as all the song and dance would soon reach the ears of the hungry predators. I guess both actions i.e. more reproduction and being hunted would have eventually balanced out and the population remained constant. The same thing must have panned out for the introverts i.e. less reproduction but increased survival ensured that the numbers remained constant. So, neither introverts nor extroverts became a majority and thus co-existed. I know it's a crazy theory but so is the human mind. I am told that we are all social animals and are bound by societal bonds. It is in our very basic genetic construct to not be able to exist in isolation. Introverts may take time off for themselves but eventually need to mingle in order to remain sane. But I believe that all people, irrespective of personality type, need time off in solitude to contemplate the bigger picture and seek the greater truth. Meditation is something that is supposed to give you immense amounts of peace and happiness, and open your mind for enlightenment. But, meditation is done alone.

It's only a play of your body and mind, in solitude. Whence then is the need for societal bonds, a curious bystander may ask.

As I seem to be blabbering on in an effort to ignite your curiosity in understanding how our society works, I'm wondering if I should share some more interesting observations. Am quite undecided, to be honest with you. Decision-making is a cause of great stress for many of us. The fear of regretting the choice and living through the consequences make us evaluate the options endlessly. So much so that we would like it if others make the choice for us so that we don't have ourselves to blame, should a need arise. Most of the good quality thought process involves decision-making. But we can't deal with it, so a vast majority of us seem to have outsourced the activity of thinking to others. We just want to be told about what needs to be done.

One of the things we don't break our heads while deciding is the chance to show off a new expensive gadget that we have purchased. While we don't go about beating drums that we bought something expensive, we do make it a point to casually slip it into people's attention. A large chunk of the human species is sufficiently intelligent but insufficiently humble. The new gadget is meant to serve a purpose for you, and let it do just that. Also, what's up with the gazillion types of smartphones everywhere these days? Yes, they are incredibly useful but let them not rob us of our everyday non-tech pleasures. Thanks to the fear of damaging our smartphones, the must-have pleasures of life like dancing in the rain are lost.

Money holds a very powerful position in society and can, on occasions, be used to compensate for immoral acts. Additionally, money also has a role to play in the pursuit of the truth we are after. Space and scientific research cost money and governments don't sponsor everything. So, a large chunk of research is funded by large private corporations. That being the case, the research then naturally becomes focused on what the large corporations are after. All said and done, we pursue the agenda of those whose money is involved and the agenda is mostly never on the same path as seeking a better understanding of the cosmos. The research is swayed by the desire of monetary profits and turns into an exercise of creating an artificial want in the market so that products can be sold. Through time, we have become increasingly familiar with the concept of forced wants and unfamiliar with the concept of basic needs.

Basic social dignity is a matter of great concern for us. We cannot be caught doing something that is not in accordance with social norms, else we run the risk of bringing great shame to our families. In the back of our minds, we believe that people don't forget. So, we start erring on the side of caution always and we become circumspect in all things related to society. It naturally starts impacting how we live our lives and people stop doing things that they should normally be doing. Fear of shame keeps a good part of our civilization's ability hidden and minds ignorant. In this day and age where information is available freely, ignorance is considered a sin. So, people want to appear knowledgeable and feel that their opinion is valued and respected. As a result, pretentiousness creeps in. Some people become very aggressive and defend their thoughts and theories like possessed priests. Two men start hating each other for arguments on topics that neither of them actually has much of a clue on. If a man feels that a theory is wrong, he will just avoid and ignore it. Not acknowledging the existence of something is one way of saying that it is wrong. Men with good oratory skills go on to assemble and preach to large groups. They impart one-way non-arguable discourses. They should know that most who preach carry the illusion of knowledge. Those who are wise and actually have some amount of real knowledge keep learning and sharing. They don't preach.

Wise babysitter: All children have natural metaphysical curiosity.
Man: But only those with a proper upbringing retain it. Some of us never had an upbringing. We just grew up.
Wise babysitter: You seem quite stressed. I have observed that you are stressed and restless all the time.
Man: I am a restless person, and cannot do anything without being stressed. That is a result of my genetic construct, my upbringing, my experiences and lessons during the formative years.
Wise babysitter: Don't blame your parents. You are conditioned by the society you live in. Constant peer pressure to excel has made you want to do everything perfectly, as a result of which action items cause a great amount of restlessness and stress in you.
Man: Your words, though true, are stressing me out.

We want to feel good all of the time. Our constant endeavor seems to be to keep looking for things that are reassuring and make us feel good. Not only good, but they should also fit into our belief system. We are biased when we search for answers, for we want them to fit into what we believe in. Anything that doesn't agree with our beliefs is either wrong or needs to be analyzed at a later point. Also, we all want to be related to those in power. It gives us the false comfort of safety and assurance. The first thing one likes to do when having a tussle with the authorities is to call up someone higher in authority so that influence can be enforced. We want to stay away from resolving the tussle by ourselves. That would mean too much thinking and hard confrontation of reasoning. The delusions of power and manipulation cloud our minds and impact sensible and practical thought.

We love magic and the mystical. We queue up for movies and shows that can take us away from our ordinary non-magical lives. We want to be fooled, yet want to have the pride of knowing that we can't be. We live our lives through the lives of fictional characters we admire. As a result, we are completely detached from the larger reality we live in. We have become conditioned that way through the society we have lived all our lives in. It is not surprising that many of us are not inspired to seek our connection with the foundation of actual reality and creation. One of the processes most essential for our creation and future existence is reproduction. But for some reason, it is a taboo subject in our society and we feel ashamed discussing it. There are norms around what is shameful and what is not. Freedom of expression is not a priority for anyone. We have created moral doctrines around decency and derive a great amount of satisfaction in punishing those who don't comply. I don't feel embarrassed to remind you that out of the many millions of unique species on this planet, human beings are the only ones who feel the need to wear clothes.

What's that? What do you mean that I don't have a heart; that I don't feel love and brotherhood of the society we live in? The heart is just an organ doing its job. The entire play of emotions is of the mind. Well, okay, I know that the last few stanzas have been more of me venting about my problems with the society I live in. I agree. I do have a problem with that. But I cannot live alone, can I? In the words of Aristotle, "He who is unable to live in society, or who has no need because he is sufficient for himself, must be either a beast or a God." I am neither and neither exists. I exist and have to live in society, and it is my role to help it improve and become a catalyst for its evolution.

I wish to travel and see this beautiful planet on which I was born and grew up. I wish to see all the majestic mountains, the endless valleys, and the great rivers. I am an earthling and should be free to freely explore every nook and corner of my planet. But those in command won't let me. I am required to have difficult to obtain permissions, complex paperwork and a lot of money. Why? I now have a very serious topic to discuss with you, and that's about the concept of nations and nationality. How did we end up dividing the world into continents and countries? The concept of continents makes some sense as they are in a way naturally occurring. Our planet consists of different and distinct landmasses separated by water. So, there is an obvious need to name and classify them so that we can be consistent when we refer to them. Fair enough. Although we should note that Europe and Asia have no waters separating them, and thus should be a single continent (Eurasia or Asirope?). But they are separate, for social and political reasons, and not because tectonic plates said so. Perhaps ancient Greeks wanted nothing to do with the odd-looking Asians.

Okay, now where did the concept of countries come from? We know that creations of nature like mountains and water bodies make natural borders for land to be divided on the basis of which. Thus, two different species can have their own private space to do their own thing. But humans are one single species. Why do we need a border? Well, we humans believe that we have subspecies within us through appearance, in the form of different races. We further believe that we have subspecies within the subspecies in the form of groups with different ideologies, beliefs, attire, and culinary preferences. So, every group needs a space of their own like how all kids need their own rooms. Such groups want to have their own rules, concepts about creation, definitions of right and wrong, and the acceptable list of moral values. All groups want to form civilizations, prosper and become better than other civilizations. They want to expand and force others to become like them and before you know it, villages turn into towns to cities to states to nations. We then have a national government, bird, animal, color, anthem, flag, and anything else we can think of that makes us unique and feel special. Creatures from one country are not freely allowed to enter the other. Each country wants to be safe and powerful and thus builds a huge army and buys weapons. Countries fight for land, resources, and supremacy. As a result of all of this, we have ended up dividing Earth into 200 countries. I like to think of this as 200 different species of human beings because that's the way we behave.

Nationality is a deep-rooted concept in our minds, possibly second only to religion. But, like religion, it is not naturally occurring. It is man-made. Every child who is born has nationality forever stamped in the mind, and that's what shapes the child's identity, ideologies, and mindset. The child grows up to display all the characteristics that a citizen of that country would, and those are different from characteristics that citizens of the neighboring country display. Thus, people of those two countries don't get along with each other. So they fight and waste time. National anthems are played before sports matches to remind us that we are divided and need to compete with each other. The concept of nationality has divided us immensely, and we are forever fighting for supremacy, for resources and to be called better and more advanced. Why do we keep forgetting that the division of land into countries is artificial? After all, we're all earthlings and part of one team. Let's remember that when we cheer for our country in the next Olympic games. Cheering for a team would only make sense if we were cheering for Earth in an interplanetary match, or for our solar system in an interstellar series or for the Milky Way in an intergalactic cup. But other teams are yet to register for such battles. We, however, seem to have satisfied our appetite for battles by fighting numerous of them on this planet. So many of them have been for the sake of countries, with millions dying in the name of patriotism. Oh, what an absolute waste of life dying for the sake of a country in a war is. We are squandering the gift to experience the magnificence of everything around us. We are losing it to fight for something artificial and irrelevant.

The great philosophers were not Greek. The space pioneers were not Russians and Americans. The great philosophers were earthlings and the space pioneers were also earthlings. We are one species on a planet. But no one seems to understand this. Men find pride in joining the army, fighting for and dying for their nations. I can't understand them nor respect such men who expect respect as they feel they are doing something for a great cause. It's a foolish cause and patriotism is a delusion. In a war, there are no heroes and no cowards; all are fools. Let's wake up, please. We have drawn borders on a piece of paper and now want to die for that one section of the paper. Also, these are not small battles we are talking of. Patriotism has led to great wars with indescribable violence and cruelty. Not only have we destroyed lives but also the untold knowledge, ideas, and revelations of millions of our fellow truth seekers. How could we let such a thing happen? Alas, there continues to be no realization and this foolish drama goes on.

Teacher: Class, let's all look at this globe of our planet. Look at the wonderful precision of the clearly defined borders of the various beautiful countries.

Wise student: You are blind. I see only one beautiful planet. I know nothing of borders. I don't belong to any land. I don't have any beliefs. I am an explorer. I am free.

Teacher: Excuse me! What was that?
Wise student: We are in a cave, not knowing why and how. Should we spend time in finding that out, and why this cave exists in the first place? Or should we divide the cave into countries and fight amongst ourselves?
Teacher: This is a geography class and not a philosophy forum.
Wise student: I am saddened by your ignorance of the geography of the reality you live in.

We are a very passionate species. Our collective effort and passion, when channeled in the right direction, can lead us to a more meaningful existence. When channeled in the wrong direction, they lead to evil acts of terror. Nationalism and all kinds of other beliefs have led to violence and brutality, which has led to hatred. That, in turn, has led to rage for revenge and fairness. All of this has culminated in terrorism. Terrorism is nothing but patriotism and religion together gone wild and blind. It takes society back to the era of barbaric rulers, with innocent civilians paying the price with their lives. It has to stop. But, terrorism cannot be ended with bullets. Terrorism is based on ideas and beliefs, both of which are bulletproof. Ideas and beliefs can only be changed through reasoning and self-realization. I believe that realization will begin when we start reasoning on what a pointless thing patriotic sentiment is. I know that such strong words may upset you, but you have to see these things for what they really are. Most brainwashed terrorists have died believing that dying for that cause would bring an eternal afterlife of happiness in the martyr's heaven, in the everlasting company of 72 beautiful and pure virgins. Seriously? The belief in a reward for dying for a cause fuels both patriotic wars and terrorism. What a waste!

What's that? You heard someone say that survival of the fittest rule applies in nature, and humans are wild by nature as they have descended from animals. So, wars are justified? Well, I think our ancestors, right from the creatures that first walked on land, have struggled enough in their lifetimes for food, shelter, sex, and survival. The survival of the fittest instinct should end with our generation. We have evolved to grow out of it. There are enough resources out there for us not to kill someone over them. There is absolutely no reason for us to fight and kill. If the point is that we just can't live without wars, then let's fight intellectual wars. The intellectual wars among priests, scientists, and philosophers on the subject of creation and existence have been more fruitful for the society, than any of the world wars. Alas, no one seems to listen; we're all asleep in ignorance.

I feel that we need a wake-up call in the form of a worldwide problem all of us can focus and work on collectively. Global warming and fear of nuclear weapons are pulling up short, and are not taken seriously. At this juncture, the cosmos needs to help us. The world needs an asteroid collision scare. The asteroid should obviously not come hit us, but only scare us into believing that it might. The other way is if we were to receive a threatening signal from an extraterrestrial life form leading us to believe that an invasion is in the offing. The euphoria surrounding the fact that aliens exist coupled with the fact that we need to save Earth would capture the minds of all people on this planet. This, or the asteroid impact scare, are the only ways I can think of where we'll all come together to think collectively and forget about our prejudiced man-made differences for a while. If the cosmos cannot do this for us, perhaps the political world leaders should fake it for the sake of the planet and its creatures. That is, lie for the greater good.

Such wish lists aside, I want to know how did our society and we humans turn out this way? Feeling no remorse in killing and causing suffering. Considering what human colonization of the various civilizations on Earth has meant for the native settlements, in the sense that it has left them destroyed and traumatized, I only wonder if human colonization of other planets is such a good idea. Our track record is abysmal. Also, if all planets had life and human beings, many wars would have already been fought by now and most planets destroyed. Apparently, in our scheme of things, no destruction means no progress. Do we consider ourselves to be an advanced species? Sure, we have harnessed the energy of the cosmos to create technology and incredible machines. But, in our mindsets, we are collectively still somewhat backward. We are very much still barbarians. It's just that we have become civilized enough to be selective in our barbaric outbursts. So, have we not progressed mentally since antiquity? We are, physically and mentally, similarly equipped as our earlier generations. It is all about how we perceive and make use of nature, and what we choose to believe in. Being advanced or backward is a choice we all make. Advancement will come through wisdom and through understanding the meaning of existence and the bigger picture. It will come through realizing our place in the cosmos and the opportunity we are gifted.

Okay, so what kind of society am I prescribing that will help us source wisdom and lead fulfilling lives pursuing the great truth about existence? Well, I don't know exactly. We need values for sure so that we have some order. But we should all know why those values need to be followed. They are not to be followed because of a fear of God's wrath or because the values are ingrained in us. Society needs to be open and empowering, where people are encouraged to become broad-minded and seek intellectual wealth. It should be a place where all the petty and useless practices like comparing, judging and boasting are wiped out. Where people are provided the ideal environment to develop wisdom on their own. Where things are corrected if discovered to be wrong.

In the context of the next chapter, I should probably mention here about what a good society should have as explanations of the working of the cosmos. A society with mystical explanations for most things and a society with explanations for all things (mystical or otherwise) are both badly constructed. A well-constructed society is the one that endorses non-mystical explanations for most things and accepts lack of capability and understanding regarding things that are not supported by solid observations and evidence. Sounds like a fairy tale?

GOD & RELIGION

Religion is the sigh of the oppressed creature, the heart of a heartless world, and the soul of soulless conditions. It is the opium of the people.

– Karl Marx

Critique of Hegel's Philosophy of Right (1843)

Light shone upon man and he said, "Let there be God." In stark similarity with how the concept of morality was created artificially in society to help humanity, so was the concept of religion and God. Yes, we created an illusion called God long ago, to help us out in answering questions for which we could find no answers. It helped us justify everything that was happening around us, including people dying and solar eclipses. It satisfied our need to know it all. Further, religion is nothing but a method of living we created to incorrectly explain the workings of the cosmos and define the values to be obeyed while existing. There, that's pretty much the summary of this chapter and all the remaining pages contain my thoughts intended to justify my claims and drive home this point. I have employed the method of 'straight talk & follow through' that I discovered in the corporate sector. So, if you are a religious person and believe in the existence of God, then you will not agree with my views. We should respect our difference of opinion. I am not here to fight a cause for you to turn into an atheist. We believe what we choose to believe in. Let's live and let live. But, if we have the power of sight, then let's not behave like we are blind.

I think of our distant ancestors living in the African lands thousands of years ago and trying to make sense of all that was happening around them. Their toil included not only trying to understand but also trying to make use of all phenomena to help them survive. An accurate analogy for this would be the advanced hard level of a first-person 3D video game where there are no clues, cheat codes, and all deaths in the game are real, including yours. How circumspect would you be if you were playing such an action game? Our fearful ancestors must have seen the Sun as a light in the sky whose presence helped them survive. They must have seen it as a living thing that came by every day and left when night approached. For them, it ought to have been a comforting and friendly face in the sky that helped them see so that they could hunt and notice any predators around. It was also a source of warmth and energy. As our ancestors traveled to colder lands, they must have wished for the Sun to be around for long so that they could keep warm. So, they started worshipping it and slowly started seeing it as a presence in the sky taking care of them and directing all that was happening on land. This belief grew stronger when they saw that the presence of the Sun helped crops grow. So that is how we invented a deity, out of the need for survival. In simple terms, we tried to leverage what we had in hand in order to survive, and that's absolutely fine. Our forefathers discovered that the cosmos they lived in was simply way too complex to comprehend, and hence they invented the divine to make life livable. If we were in their place, we would have done the same thing as that made absolute sense at that point, and we knew no better. A belief in God helped us to keep going and exploring. A belief in deities subsequently led to religion and that answered all the questions about creation, death, afterlife, nature, and values. So, it satisfied our curiosity and also helped us deal with sorrow. It helped us make sense of things and get by. Those happen to be exactly the same as the pursuits of our life right now i.e. make sense and get by.

Okay, so that's how the roots of this tree of religion and God were implanted. But how did this grow into a forest that went on to extend to all parts of the planet and our minds? Well, it's a very long story. As we have come to realize, we have a soft corner for stories. We also have the gift of imagination and creativity. Men used these gifts and created religious tales full of wondrous creatures, gods, and everything imaginable. You should know that we humans have a natural tendency to create local versions of anything and everything that we come across in our travels and are impressed by. Our imperfect tendencies of plagiarism also played a part. Cultures created their own versions of how creation happened and associated magical stories. The ship of storytelling helped religion sail across the great waters. These tales moved on from one civilization to another. In some cultures, they believed that the world was a level field between two mountains that held the sky, and stars were lamps hanging from that sky. Such tales and the religious doctrine dovetailed nicely with all the questions that were puzzling people all around, and thus did not face any resistance. Those who resisted were executed. The concept of God came in so many versions and was so easily applied to so many phenomena, that it became impossible to disprove it. Also, there was something mystical about the whole thing, so that went in favor of it for we are always attracted to anything that's out of the ordinary. People saw darkness as an evil force to which the Sun would lose the battle every evening but then defeat every morning. They saw faces in the sky and in the mountains. They saw spirits roaming around. Everything was through divine design.

Men started naming everything after the gods and creatures from these tales. And by everything, I mean every single thing. The Icarian sea in Greece is named after character Icarus from Greek mythology, who unfortunately fell into that sea after he flew too close to the Sun and burnt his wings. Yes, one should not fly too close to the Sun. With time, random events kept happening that solidified our belief in a supernatural almighty. We have already harped more than enough on how things happen randomly in our cosmos. In our society, people who lead healthy and value-driven lives sometimes die due to sudden illness or accidents, for no fault of their own. Children are born with birth defects. This is all random. This randomness also applies when someone who is on the verge of dying somehow survives. But, people see that as an act of God or divine intervention. We believe what we choose to believe in. With such random events being seen as divine miracles, the existence of God started slowly becoming a brute fact.

In a moment of great pain and sorrow, you may turn to calming narcotics or to comforting faith. Both are equally dangerous.

The spread of the concept of divinity was also helped by our emotional weaknesses. You see, we are not emotionally strong enough to deal with the death of a loved one. The sorrow and trauma are just too overwhelming and one can lose the will to live. The concept of God has helped humanity cope up with untimely death of loved ones, justifying it as fate and God's will. It has told us that there are things to which we must all surrender; and that the dead go to a better place and spirits live on. These are all meant as coping mechanisms so that the lives of the mourning people can go on. Please take a moment and think about it. A young girl whose father dies can be told that her daddy is waiting for her in heaven, and he can hear what she prays for. Or, she can be told the truth about death and that her daddy is forever gone. What would you tell a little girl?

Hope, fear, and ignorance are things that are very natural to us and therefore make us suckers for anything supernatural. Our awestruck ancestors saw patterns in the stars in the night sky and imagined them to be mythological creatures. They gave them names and started worshipping them. They connected all dots they could find and everything was related to God. Any new thing they came across whose origin was unknown was immediately identified to be of supernatural origin. Naturally, some things and events earned the tag of good and some became forever evil. Our clueless ancestors thought illnesses were caused by evil spirits. Earthquakes and floods were caused by demons. They observed that some actions helped them achieve desired outcomes and thus religious superstitions were born. I can go on and on about how this whole opus unfolded. As we can see, one thing led to another and, lo and behold, religion was everywhere and in everything. Once people were convinced that God made everything, then it was only a case of looking at things and coming up with stories on why God must have made it that way. There was no longer a distinction between fact and fantasy.

Religion advocates that you control your life by surrendering control.

One can say that all of this happened in the eras of great ignorance, so such thinking and beliefs can be understood and accommodated. We live in enlightened times now, but the thinking and beliefs of most of us remain the same. We now know what the Sun is. It is the star of our solar system. There are hundreds of billions of them in our galaxy and about a billion trillion of them in the universe. They are celestial entities that are formed through matter obeying physical laws, and they die when the fuel inside them runs out. We know all of this and a whole lot more about them. But we have retained the beliefs our distant ancestors had and still worship the Sun. Why didn't the advancement of the human civilization, including a thorough scientific understanding of the cosmos, enable us to do away with such beliefs? We can keep picking our brains but alas, hundreds of millions of us still worship the Sun as a benevolent deity and seek blessings, which makes no sense at all. The Sun is not a conscious being with a mind to realize what's going on among tiny fragile beings on a distant planet and tell them to stop this practice. We are supposed to have a mind and intelligence to not be indulging in activities like these. Anyone listening?

Religion is the refined sugar of our times. It is in everything we consume. You may see the harms of it but the society is not going to stop consuming it.

You know my name as it is on the cover of this book. From my name, you can figure out my religion. Nearly all of us are named after characters from religious tales. We are born into a religion, instilled beliefs based on that religion and we live our lives adhering to those. Religion has turned into such an important part of our lives that it defines everything that we do. It dictates everything, right from my upbringing, my definition of right and wrong, my attire, my cuisine, to how my body is disposed of after my death. It has taken the form of our fundamental understanding of reality. With time, various different religions have sprung up and evolved, each talking about their understanding of everything and preaching their beliefs. We keep forgetting that all of this is man-made i.e. the concept of religion, those religious tales and the characters in them, and the laid down beliefs and principles. But, you are naturally occurring. So, why is everything that you believe in then artificial?

When choosing which religion to embrace or which God to worship, the choice does not matter. The wrong choice has already been made.

We have mistaken something downright artificial for something that's not only naturally occurring but the most fundamental natural thing. Millions of men have spent billions of hours in studying, contemplating, writing, revising, teaching, preaching, and philosophizing everything related to religion. As a result of all of this, we have hundreds of massive sacred scriptures in various versions and languages. They tell tales about gods and their acts, about morals and the supernatural. They are like 'how to live your life' guidebooks and are fully illustrated with stories that we are meant to believe are real-life incidents. We are forever engulfed in these books, and find no time to read the book of reality all around us. Let us please remove the blindfold of religion and read the book of reality. The book of Genesis belongs in the fiction section of the library, and so do all the other religious scriptures.

Preachers of religion have a great sense of peace and calm around them, of apparent contentment arrived at after knowing all the answers. A heavily sedated mad man also has a sense of peace and calm around him.

Wise parrot: What or who created the universe?
Priest: God created the universe.
Wise parrot: And who created God?
Priest: God is all-powerful, omnipresent and eternal. He has the ability to summon himself into existence out of nothing. He is the absolute.
Wise parrot: If God has a human form, which is what we see in most sacred places and holy books, then why was the first life on Earth not human? Why did evolution need to take place for billions of years for humans to emerge?
Priest: That's because God intended it that way. He created and guided the evolutionary process.
Wise parrot: Are you insane?
Priest: No. I am the way God made me, and so are you.
Wise parrot: It is pointless discussing with you.
Priest: Do you have an alternate explanation for who created the universe?
Wise parrot: Not completely, but there was definitely no intelligent conscious creator. The science community is working on finding the exact answers.
Priest: If there was no creator, then what caused the phenomenon that created the universe?
Wise parrot: As I said, that puzzle is yet to be solved. Let me ask you this: what or who created God, for him to then, in turn, create the universe? Up in the heavens, God must wonder: where did I come from?
Priest: I already told you that God has the ability to summon himself into existence out of nothing.
Wise parrot: I pray to your God that you be summoned into non-existence this very moment.

The masses choose to remain blinded and forever drunk on the supernatural. Ready to believe without reasoning.

We have a very strong feeling of excitement and attraction for anything that is surreal. That includes anything that can give us an experience of transcendence. We search for it in sports through men who can achieve unparalleled superhuman feats. We want to look up to and worship such men. We search for it in music through tunes that can put us in trance. We want to be blown away by ethereal works of art. It is not surprising then that the surreal concept of God has captured our imagination and settled down to become a part of us. Our innate craving has led to the creation of an endless supply of beliefs so that our minds feel satisfied.

The mystical carries its own gravitational pull.

The mindset of parents and the environment that exists in the household define a child's upbringing. I was raised in a conforming household. We prayed to God to answer our prayers and celebrated God's existence and triumph through festivals. The word doing the rounds was that everyone needed to follow dharma and live the right way by following virtues. As a child, I would go to the temple to join my hands and pray. I would pray for some wishes to be granted and for good things to happen to my family. I went through the motions, doing things as instructed like an obedient child. But, questions of reality were starting to emerge in the suburbs of my mind. As time passed and I experienced the workings of the world around me, I slowly started realizing the actual truth. I came to realize that prayer is one-way communication. Religious places contain inanimate idols made of stone. That is the plain reality.

Everything else that happens there is fantasy that goes on in the mind.

Prayer is the tonic that kills the brain cells in charge of reasoning and logic.

The only good thing about prayer I guess is that it gives you a brief meditative phase to collate your thoughts. It gives you some clarity on what exactly you are seeking at that point in your life. That's about that. But people don't do it for those purposes. People do it to ask for things, to feel good as they feel that it is time well spent and will ensure good returns in life. Also, prayer only happens when you have faith, which is nothing but the trust you have in God or the religious beliefs. I've heard people saying that praying keeps them grounded. Having faith makes them humble and human. Well, you don't need to carry out an elaborate invocation act in order to feel human. That's who you are. You just need to use your common sense and get the job done. Praying for it won't get the job done. Look at it this way: If you want to cook a meal, you must source the raw materials. Then you need to prepare it as per the recipe. How well the meal turns out depends on the raw materials, recipe and your skills. This is called common sense. Staring at the raw materials and praying for them to self-assemble into a tasty meal is called faith. I know that I am sounding pretty blunt here but I did give you a heads-up that I would talk straight and not beat around the bush.

Faith is an obsessive disorder, without a medicinal cure.

Riddle me this: Out of all the 9 million species of creatures that currently exist on this planet, why do only humans beings feel the need to pray to God? There is absolutely no evidence that any non-human species believes in God, prays, does rituals or has any religious beliefs. None whatsoever. Are you saying that animals lack the mental acumen to realize God? That's rubbish. All animals are intelligent enough for they understand things, know how to strategize and hunt, how to carry on their genes and what not. So, what does that mean? Are humans only the chosen ones by God to have the privilege of knowing its existence? That would be very unfair and discriminating. Or, is it the case that 1 species is right and 8,999,999 species are fools? The pigeon in my balcony and I wish to know. Tell us, please.

God is a myth, like a flat Earth. Such myths were once necessary. But are now considered errors in judgment.

The evolutionary wild instinct in us seems to have found a way to escape extinction and a channel through religion to remain alive. From barbaric rituals to fighting wars with humans having a different faith, we have gone all out in doing countless inhuman acts in the name of God and religion. We have divided ourselves, started hating and hurting each other. What for? For something that does not even exist? This is just wrong; so very wrong. History tells us that people who questioned and opposed the existence of God were ridiculed, tortured and killed. Innumerable innocent intellectually driven people were killed by religion for thousands of years. They were executed because they sinned by choosing to imagine the true picture of the cosmos in which our world was one of many, created by physical laws. They erred by questioning the existence of a creator. Religious leaders who committed such hideous crimes lived a life of happiness, worshipped by their followers and being blessed, in their eyes, by their creator.

Shouldn't we question who eventually paid for these crimes or sins? Apparently, no one did. These cases await a trial.

In a land where all have always been blind, the one born with sight will be proved a fool.

Widely considered as the father of modern science, the Italian astronomer and physicist Galileo Galilei used the engineering marvel called telescope to help us realize that the Earth revolved around the Sun. He told us that the laws of nature were mathematical. He told us things we never knew about the planets. His contributions were like none other in the past. The church took great offense to his findings and inferences as those violated sacred scripts. He was tried by the church in 1633 on charges of heresy and forced to recant. He had to spend the rest of his life under house arrest. The mob of the blind fools had their say.

Religion has the ability to make you commit a horrendous act and yet make you believe that you have done a righteous deed. It will also find you many supporters for the same.

Let me also tell you the story of Anne Askew, who was born in 1521 in England. She grew up to be an educated, brave and intelligent woman who held very strong beliefs. She became a writer and a poet. She was a devout protestant and went on to question the teachings of the Roman Catholic Church. She formed a group and preached about what she felt was true, and this obviously led to a lot of controversies and her being arrested. She was charged with heresy and was tortured. She was pressurized to renounce her beliefs and disclose the names of her like-minded women friends, but she did not budge. Such was the courage of this woman and her belief in the truth. The church had her tortured brutally at the Tower of London, including having her body stretched apart till her shoulders and hips were pulled from their sockets. She was carried in a chair and burnt alive publicly at the stake on 16th July 1546, at the age of 24. The burning was a well-attended affair like all other executions and a source of entertainment for the masses. What did Anne Askew do to deserve this horrific death? She did not cause any harm to any human being. She simply stood her ground on what she believed in, which was to question some beliefs of the church that were artificial and foolish in the first place. But the bishops of the holy church and the 'omnipresent benevolent God' dealt her this brutality. I have nothing to say.

It appears that the brains of most humans have just that precise amount of intelligence to not be intelligent enough to make use of that intelligence.

The Polish writer Joseph Conrad once said, "Being a woman is a terribly difficult trade since it consists principally of dealings with men." There is little doubt that women have been subject to ill-treatment since antiquity. As noted in the chapter on evolution, we have seen how women took up the role of housewives as societies took shape. Once the societies took rigid shape, women had more misfortune in store for them. I would like to list out a few here since we need constant reminding of our inglorious past:

- Female child infanticide
- Ritual widow murder practices like Sati
- Genetic mutilation
- Treatment like slaves, subject to all forms of abuse
- Highly unequal civil rights and education opportunities

Why this injustice against women? Purely because religion said so. I'll not name the religions but rather summarize some of the things they preach in various sacred scriptures:

- The value of a woman's life is around half of that of a man's life.
- Women are morally and intellectually inferior to men. Their main job is to serve men sexually and in housekeeping.
- They are to be forever dependent on their fathers or husbands and are to do things only upon their permission.
- Their life has no purpose if their husbands die.
- Disobedient wives should be beaten.
- Their primary purpose is to give birth to male children.
- Women sexuality is a taboo matter, and never to be discussed.

Hindu families worship female gods like Lakshmi, but in their minds they always want the newborn child to be a boy. The Christian Old Testament reads that God says that if a woman does not scream loudly enough while being raped in the city, she should be stoned to death. All this is what religion teaches men about women. Religious men then carry out God's mandates to the tee, while wicked religious men add their wild barbaric instincts in the mix and do all kinds of sickening acts. If this is what religion thinks of half of the human species, then does it deserve a place in this precious world? Go figure!

I cannot talk of religious crimes and not mention the gruesome act of human sacrifice that's been happening in civilizations all across since I don't know when. Scriptures talk about sacrifices as virtuous acts. The biggest story in the Bible is that God sacrifices his son, Jesus Christ, for the sins of mankind. Thus, men have committed such acts simply because the gods did it.

- The Mayans sacrificed children, including infants, to satisfy the supernatural beings.
- The Aztecs sacrificed humans, including children, every day to please the God of rain.
- The Incas sacrificed children through Capacocha rituals in order to stop natural disasters from happening.

Considering the vast length of the reign of these civilizations, I cannot even imagine how many innocent children would have been killed in the name of God. These are just three of the various cultures that committed such shocking acts on account of sheer human stupidity. No human can have the right to rob innocent children of their lives, but religion allows it freely. As nearly as about 200 years ago, a boy was killed every day at the Kali temple in Kolkata in order to please the goddess Kali. If this is the effect that religion has on our minds, then what good is it I ask? Although barbaric practices like child sacrifice are now banned, we continue to make offerings to God through other forms in order to please. In simpler words, we continue to remain stupid by choice. I am not even going to mention the savage practice of animal sacrifice, which has been going on for eons. It appears that animals have no rights whatsoever in the place they call home. It seems not a long time ago that we were discussing rich topics like ways and means to figure out if other life exists in the cosmos. And here we are now discussing such lowly and shameful acts of humanity. Do we even deserve a special place in the cosmos?

Religion has prevented the world from being a blessed place.

We feel that our sins can be washed, blessings gained and spiritual enlightenment obtained by participating in religious pilgrimages at holy places. At the Kumbh Mela in 2019, close to

250 million devotees visited! Now that calls for some serious event management infrastructure, which is quite impossible to do. As a result, there are tragic accidents. In one such accident, around 2,500 pilgrims died in a stampede during the annual 2015 Hajj pilgrimage in Mecca. That's just one incident. Thousands of people, including children, die frequently in stampedes during religious pilgrimages all over the world. Why would God take the life of someone who arrives with noble intentions? What a waste of life in the name of worship! This is as idiotic as brainless mass cult suicides. But some remarkable humans have found reason in this as well, stating it as the intention of God to meet with the people who died in the process of worship, and those dead are with God as per his will. This makes me reword a famous movie quote as follows: "*I see stupid people. They are everywhere. They only see what they want to see. They don't know that they are crackpots.*" When will such dim-witted scholars realize that you don't need to die in order to meet your maker. Your maker is nature and it is all around you all the time.

In the previous chapter, we had noted how nationalism has led to division in society and extreme violence. It has also led to terrorism after we added religion in the mix. I have no doubt in my mind that religion and nationalism are the greatest killers on the planet, and both of them have jointly caused all the terrorist activities in the history of our world. I have witnessed religious riots first-hand, and have seen the brutal hatred in the eyes of people for fellow human beings of a different religion. I don't think anyone has ever tried to calculate the total number of deaths on account of religious hatred and wars. It should be a gargantuan number, no doubt. Let's not waste time on such disturbing math. In the context of normal society workings, nationality and religion are the parameters on which judgment on people is passed. If we hear that a person is a Jew from

Israel, then we already have a mental picture of the appearance, mannerisms, beliefs, and quirks of that person. In other words, we already pass judgment on that person purely on the basis of nationality and religion. Incredible. We do indeed deserve a very special place in the high heavens for such noble acts.

Man: Did you hear about the discovery about the God particle? We now have scientific proof of the existence of God.
Wise tea-seller: I don't think they have found any proof of God's existence. The Higgs boson or God particle is an elementary particle, in the realm of science and not religion.
Man: Oh, is it? My bad. I guess God is going to make us wait a bit more before he appears. We need to continue walking his path.
Wise tea-seller: For something to appear, it has to exist. God is anti-science fiction.
Man: Where do you think all this complexity around us came from?
Wise tea-seller: Yes, complexity means there is design involved. I agree that there is an intelligent creator behind the making of man. The intelligent creator is evolution. It is so intelligent that it does not need to prove its existence. Just like God.
Man: Your sarcasm does not amuse me. I believe in what I feel is right.
Wise tea-seller: People believe what they want to believe. From seeing light in near-death experiences, to out of body experiences, to being visited by spirits—a weak mind will succumb to foolish imagination and see it as reality.
Man: I am ignoring whatever you are saying.
Wise tea-seller: Ignorance is short-term happiness.

You can only doubt what someone else claims to know.

In the past, I have argued with myself that the teachings of religion help us become good human beings and lead fulfilling moral lives. Those who commit crimes in the name of God have wrongly interpreted the message of God and are at fault. The concept of God is not at fault and we cannot stop believing in it because of a few sinners. It would be like stopping all scientific research and advancement because of incidents like Hiroshima and Chernobyl. That would, of course, not be right. While that argument holds some weight, I have realized with time that there is a fundamental flaw in the foundation of God and religion. Namely, that it is not true. You would agree with me that our lives cannot be a lie. It's like believing in the illusion of card magic because it teaches us good values. Once we realize that it is an illusion and why values are required in society, we should become good human beings because we want to. Because we want to lead good lives doing what we are interested in. That is, we should do things because of proper reasoning. It does not require fear of a God punishing you for you to be doing good deeds. You can be kind and work collaboratively, just on your own. Additionally, if people are so frequently misunderstanding the teachings of God and committing dreadful crimes, then why run that risk at all in the first place? Our minds have sufficiently evolved to handle the truth, which is that there is no God. What you have been hearing a lot these days in religious debates is true, which is that there is no supernatural deity taking care of us and blessing us all. In a world with a God watching over things, surely there wouldn't be things like childhood cancer and sexual abuse of children. Additionally, an intelligent and benevolent creator would not have entertained any worship of it or allowed wars between groups divided by their beliefs in different versions of it. All of this leads us to the divine inference that the divine does not exist.

The creator is always more intelligent and complex than the creation. In that context, I wish to meet the creator of God.

As it turns out, theism is the norm and anything contrary is the exception. An atheist apparently is a person who does not believe in the existence of a thing whose existence has never been proven. Doesn't it sound vague? An atheist has to wake up every morning and remind himself that nothing imaginary exists. Thanks to how things have turned out, the concept of the imaginary is all too real, and all other concepts are considered imaginary. Ordinary claims are being asked to be backed up with extraordinary evidence, while extraordinary claims are considered self-proven. In an ideal world, there would be no atheists out to violently prove the non-existence of something that's non-existent. It's a no-brainer like none else. I am sure that you have already heard of the convincing atheism argument from super-intelligent men like Hitchens, Harris, Dawkins, and Dennett ('fantatheistic' four?). If you haven't, you should make time to do so.

My average yet wild intelligence makes me think of it this way: because religion makes a person blind towards the real truth about creation and existence, it has gradually made most of the world blind. But this was not an intentional and manipulative act in the beginning. It just happened that way. The illness of blindness spread uncontrollably from one person to the other. None of the blind thought that they were blind. There is no authority that forcibly blinded the masses. It is the blind that formed the authority so that they could reaffirm their beliefs and reassure meaning in their own existence. It is in our nature to make people believe in what we believe in. We find that comforting. So, that's how the belief and blindness spread. Of late, this has sadly become intentional and manipulative.

People and institutions are leveraging the masses' belief in God to reap monetary profits. Religious pilgrimages are million-dollar businesses. Conmen gurus claiming visions of God have sprung up everywhere looting the minds, dignity and bank accounts of foolish followers. In its own evolutionary cycle, religion started as a concept, then became a solution, then like a guide, recently turned into a business and is now an outright scam.

Every generation turns out this God worshipping way because of the upbringing of all the children. In our society, nearly all children in all religions are taught about the existence of God and the values derived from religious scriptures. Children believe everything parents and elders tell them. As we have learnt, it is very difficult to unlearn what one is taught in childhood, especially something as fundamental as the nature of reality and origin of things. If you don't teach your children about God, they will never ask you that question. They will ask you questions about the night sky and creation. You can answer those with science and reality, or with religion and illusion. We are guilty of brainwashing children with a fictional view of reality. A child may have the best upbringing in terms of facilities and resources, but if you instill false concepts like heaven & hell, then the upbringing is not a very good one. We should use religious tales judiciously as fictional stories to instill good values in children, who can then go on to lead a value-driven life pursuing everything that is real. I feel that every child is born thinking like a metaphysician, toddles around playing the role of a theoretical physicist, looks for things to become an experimental scientist, but then is brainwashed by society to think like a priest. Let's remember that till children are exposed to the outside world, they will believe everything we tell them. The foundation of their lifelong thought process is based on what we teach them. Let's teach them the right things so that there is not so much difficult unlearning they need to do in the future and embarrassing realization they need to face of the fact that their parents were fools teaching the wrong concepts.

"Good night, sleep tight, wake up bright, in the morning light, to do the right, with all your might." This is the prayer I used to recite every night before hitting the sack, during a particular phase of my childhood. I had learnt this poem by watching television. As it turns out, the religious upbringing by parents is further enforced in life by the entertainment media. Once you grow up a little and have access to television, you'll find that the teachings of scriptures are broadcast subtly through movies, soap operas, and songs. Good always prevails over evil, and the sinning bad guys always get caught. Movie plots remind us that we should be doing good deeds, and no sin will go unpunished in the long run. We know that this is not true, on account of the real-life examples I have stated in the previous chapter. Anyway, religion and its teachings are everywhere to be seen once you open your eyes upon waking up. In India, even if you don't open your eyes you can hear them through loudspeakers. The net result is that it is the pond in which we find ourselves and are forever swimming in. We need to get out of it in order to reach the shores of the great ocean of real truth.

In a way, religion is what takes shape when beliefs are allowed to run amok. I find similarities in some professions that have evolved into a way of life with their own set of codes and ideals. One such is Bushido, which was the Japanese code of conduct and way of life for the warrior clan called Samurai. This military community observed a code of moral principles, which defined their lives and also complemented their violent existence with patience and serenity. There were eight core virtues of Bushido: loyalty, righteousness, courage, honesty, benevolence, respect, honor, and self-control. These had evolved over the course of four centuries and had elements of some eastern religions in them. For the Samurai, these were a sacrosanct set of values that had to be followed no matter what, and their purpose of existence revolved around them. They lived on beliefs like all decisions were to be made within the space of seven breaths. They saw fighting as a spiritual exercise and had no fear of death, often committing ritual suicide in order to avoid capture. They lived their lives like possessed and programmed machines. Bushido is an example of what values and beliefs gone wild can do to a profession.

Man: Oh wow, how the hell did you do that? I haven't seen a wizard like you.
Wise magician: There are no wizards. There are only magicians with tricks.
Man: I guess you are right. Gods are the only wizards. We humans are not.
Wise magician: God is like magic. He does not exist.
Man: What did you just say?
Wise magician: Religion takes the world into darkness, fooling it to believe that it is light.
Man: I am offended by your views for I worship God. He is the one who gives you the skills to perform magic and earn your living. Why do you not believe in the almighty? Are you an atheist?
Wise magician: Why do I need to waste my time explaining to you if and why I am an atheist? By dwelling into the very subject, we are making a mockery of our intelligence. We ought to have better things to do in our lives than keep debating on what's real and what is not.
Man: God is real and whatever you say does not matter. You are a fool.

Wise magician: There is nothing more exhausting than talking practical sense into a religious fool. Running an ultramarathon is comparatively easier.

For a hungry man, science gives a meal that doesn't taste great but is very nutritious. Religion offers a readily available tasty meal devoid of any nutrients. The starving man often goes for the latter.

Once your mind strongly believes that something is real, it becomes real in reality for you. Nothing anyone says and does can then change that reality for you. For example, if you believe it's true that this page is painted red in color, then it will always be red for you. Nothing anyone says will change what you see. This holds true not only for people's belief in God but also for associated beliefs like they can become virtuous by visiting a place of worship and praying, for they have invested time and effort by visiting and hence showed their obedience to God. Thus, good things should naturally happen. There is also a feel-good factor associated with any act of worship, which leads to such thoughts. You tend to get into a good mood and thus positive thoughts ensue. This has all got to do with conditioning. We are brought up this way so we are in a way programmed to think in this manner always. We have clearly lost the plot in terms of everything associated with praying. A praying mouse can never escape from a preying eagle. The only good part about praying, as we had discussed, is that it gives you some moments in solitude to collate your thoughts and needs at any given point in time. Further, to meet this need, we have a billion-dollar music industry focused on creating devotional music. We know that music can lift our spirits and move us. Tunes of Sufi music are so beautiful and calming that one would want to believe that everything associated with them is true. Phrases like "*Fanoos bankar jiski hifaazat hawa kare, woh shama kya bujhe jise raushan Khuda kare* (that which, as a lantern, the wind protects, how can this flame die out that glows by the will of God himself)" are so deep that one would want to forever be lost in its mystical translation. There is little doubt that devotional chanting music, accompanied by trained voices, is a kind of therapy and is ideal for reaching peaceful meditative states. Those states are ideal for studying, writing, contemplating mysteries or doing anything that needs focus, religious or otherwise. We need to give the credit to prayer for bestowing us with this comfort. The need for prayers also leads to the construction of places of worship. We find them everywhere. These act as temporary abodes for some peace of mind, focus and escape from everyday stresses. But, mile-long queues to have a paid glimpse of the idols and then head back home in the chaos defeat this purpose completely.

Faith makes you rigid. It hardens the natural free form of your mind.

Some moral men have argued with me that the past dark ages were times when we were largely ignorant and uncivilized. That being so, the value of human life was considered to be of much lesser value than it is today. Thus, killings and sacrifices in the name of religion have to be seen in that light. My counter-argument has been that if people in the past were so ignorant and uncivilized, then why have we taken their writings and teachings of religion so seriously? Why have we chosen to believe in what they believed in? Further, why should we base our complete life and beliefs on holy scriptures that were written a couple of thousand years ago by ignorant men? It probably applied to the understanding then, in line with the universe revolving around the Earth, which sat on top of a turtle. It, for sure, does not apply to our understanding now. Those scriptures need to be done away with. I wonder why persistence with foolish thought, even when educated otherwise, is not considered a sin and appropriate punishment isn't given.

The creator is not godly. It is the creation that is godly.

Mbombo, also referred to as Bumba, is the creator God in the religion practiced by the Kuba people in Congo, Central Africa. Those people believe that in the beginning, Mbombo lived alone in the darkness and the world was full of water. He suddenly felt an intense pain in his stomach and subsequently vomited the Sun, Moon and the stars. The heat from the Sun evaporated the water on Earth, leading to clouds. Hills emerged from the water as more time passed. Mbombo vomited again, and this time out came nine animals (leopard, goat, scarab, leopard-like animal, fish, crocodile, eagle, tortoise, and a heron) and many men. They went on to somehow create all creatures in the world. Quite a story, that one. Such interesting creation stories are part of many religions, which talk about men arriving in the world in their full-bodied forms. That is, to hell with natural evolution and all that razzmatazz. We were simply constructed in our current fully evolved forms and that's how we've always been. I am interested in knowing what various religions believe in when it comes to the origin of man.

- The Genesis narrative in Christianity says that the first man, Adam, was created by God from dust. It also says that God created all of the heavens and the Earth in six days.
- The Quran in Islam talks about the first man and woman being made from clay and brought to life by the divine breath of God.

- As per the Agganna Sutta in Buddhism, humans originated in a higher form, having godlike characteristics, and slowly deteriorated into human forms.
- Jainism states that humans have always existed. There is no origin point.
- Hinduism has various different gods (no one knows how many in total) and a unique and consistent creation story is difficult to be arrived at. Although I am very interested in the story that the river Ganges was released from God Shiva's hair in order to serve the needs of the people.

Amidst all of this religion bashing and blasphemy that I seem to have completely and sinfully immersed myself in, it is time to take a break and explore the origins of something interesting. I am interested in knowing as to how the hell we ended up determining the personalities of people on the basis of their date of birth. In other words, we need to solve the mystery of zodiac signs or astrology (not to be confused with the science astronomy). Astrology played the role of astronomy back in the day. It was the method of studying the sky, recording observations and capturing them in a structured and systematic way that made sense. That's absolutely fine as it helped us to be organized in our research. It took the wrong road when we started connecting the outcome of events on Earth to the placement of celestial objects in the sky, and especially the connection of placement of planets at the time of someone's birth. Think of it as a supernatural forecast that we could rely on. It started in the Babylonian era 2,000 years ago and became more complex as it spread all across. For the life of me, I cannot understand why many of us believe in astrology. How can the constellation in which planets were at the time of my birth predict or affect my future? Isn't this downright stupid? In some remarkable cultures, horoscopes are used for wedding matchmaking. An expert can apparently read my palm and predict the rest of my life based on that. I rest my case.

The world of fantasy does not exist. And its non-existence makes the real world authentic.

Pascal's Wager is an interesting philosophical argument that presents mathematical reasoning derived benefit in the belief in God. It states that if one believes in God, the returns one can expect are far greater than if one doesn't. The basic statement is that if you believe in God and if he exists, then you can expect eternal happiness through heaven. If he doesn't exist, then nothing is gained or lost. On the contrary, if you don't believe in God and he actually exists, then be prepared for eternal hell. While the whole premise is fantasy, the net advice presented is that it is more beneficial to believe in God. I find some similarities to this when I see that some religions preach and many men believe in a formless form of God, like a single source of energy that is beyond any structure. It is the all-powerful, invisible, and unimaginable force that sees and guides everything. The Vedas, which Hindus consider to be authorless and out of this world, define God as a single formless entity. By not giving God a form, does it lessen or enhance the divinity? In my mind, it does not make a difference. At the root of it, the concept of supernatural still remains. It's just that it does not have a form, so there is no mental image that one forms when one worships. The fallacy still remains, albeit in a new avatar that adds to the mystique. The sinning blasphemer in me sees the people who worship a formless form of God as taking the conservative path. One part of their mind tells them that there is no God, but the other part convinces them to still pray to something invisible just to play it safe and be in the good books of the almighty, so to speak. So, if and when one day God shows up, they will not be punished by him for not believing. I, for one, will be straightaway banished to hell when that day comes.

Painter: Congratulations on your safe return from the successful space mission. I can't imagine the things you must have seen.
Wise astronaut: Thank you. You know, you see those images on TV to get a feel of outer space but you can truly experience it only by being out there.
Painter: No doubt. What I would pay to be out there and paint on God's blank canvas!
Wise astronaut: No creative artists' inspiration is complete without experiencing the beauty and harmony of deep space.
Painter: I pray to God to uplift me and give me a chance to experience the beauty of that unfathomable darkness.
Wise astronaut: After having seen how many people asked me if I saw any UFOs in my voyage, it is not hard to imagine why they believe in a supernatural creator. Fantasy is all too real for many.
Painter: Without the supernatural, what good is my existence as a painter? I believe that the universe actually does not exist in reality. It is just a thought of a divine mind. It is a thought in which we and everything else exists.
Wise astronaut: I am charmed by what you believe in. Your artistic creativity extends to your thoughts as well.
Painter: We're all ignorant. Some more than others.
Wise astronaut: Some way more than others. When I met the gods, I asked them on why did they make all these other worlds for nothing? Why did they bless only Earth with life?

Painter: I can sense your increasing sarcasm. Don't ridicule what you don't understand. Be humble enough to ask questions in order to gain an understanding.
Wise astronaut: When seeking something unreal, all questions are foolish and irrelevant.

In the dictionary of the realists, there exist no words called faith & luck.

Don't even get me started on the concept of luck. Phrases like 'good luck' and 'all the best' have entered all layers of our consciousness and we cannot complete a conversation without them. What is luck supposed to do? Sway the outcome of an event in your favor? Rubbish. The outcome of an event depends on the actions in the event and if there is an equal probability of it going either way, then it is down to pure randomness. Nothing can turn it one way or the other. Wearing certain ornaments and doing rituals will not help swing the outcome in your favor. Nothing brings good luck or bad luck. You make your own luck by using common sense and ensuring that the right causes take place so that the desired effects can be obtained. It is high time we stopped thinking of luck as some supernatural occurrence that can be called on by acts of ours. Alas, I forget that we live among those who believe in spirits, demons, dragons, and flying monkeys. Good luck to me.

Anything that overpowers logic and reason cannot be good.

In the profit-oriented and selectively moral corporate industry, companies follow some kind of sacrosanct religious doctrines of their own, in the form of organizational policies. Created by HR in conjunction with the leadership team, the organizational policies establish the rules of conduct and apply to everyone without exception. They ensure that the rights of employees are protected while also taking into consideration the business interests of the company. They list out the guidelines for workplace conduct and punishment for non-adherence, to be enforced objectively without any biased discretion. Policies are cited when people are in doubt, and the course of action is decided accordingly. From this analogy, it is not difficult to figure out why religious doctrine is required in society to maintain order. We are simply not structured enough to live peacefully doing the right thing unless there is a fear of punishment. We should evolve faster to get out of this. Can we though, for we are brainwashed with the notion of guilt. Religion thrives on the feeling of guilt. It is the lingering fear that we are preoccupied with on whether we have done the right thing or not. It leads us to stress on decisions, weighing the moral pros and cons. It also causes us to stress over decisions that have already been taken. Religious guilt can make life miserable. Nobody wants to have the feeling that something miserable is going to happen because they didn't do something in accordance with the holy teachings.

In the court of reasoning, the existence of God has not a single witness in the history of time. Thus, a fair judgment must be passed.

Religious philosophical thoughts have consumed the minds of many great thinkers, who ideally should have utilized their intellectual prowess to mull over science related philosophy. But, they have spent their lives engrossed with the philosophy of religion. One such is the *a priori* ontological argument to prove the existence of God. Back in the 11th century, St. Anselm of Canterbury came up with his discourse *Proslogion* in which he claimed to derive the existence of God by employing a very sophisticated deductive hypothesis, which I have attempted to simplify as follows:

1. X = A thing that exists only in your mind (imaginary thing)
2. Y = A thing that exists both in your mind and in reality (real thing)
3. X & Y together make up all things in the world
4. Y is better than X
5. God = a being than which none greater can be conceived
6. If God = X, then statement #5 is incorrect on account of statement #4
7. Thus, God = Y
8. God exists

As you would expect, such paradoxical proofs are not falsifiable but are quite interesting to read and analyze. If only these scholars would have taken up the philosophy of true science, we would have progressed so much more as a species. The process of arriving at the right answer by eliminating the negative hypothesis in an iterative manner does not work when dealing with fairy tales.

Just wonder is not what counts. It needs to be an open and genuine wonder. I can wonder about how was the world made, or I could wonder about how God made the world.

I don't know about you but I feel that we are so backward as a species. We barely know what we are ignorant about and are still consumed in arguments over whether God exists. We are wasting precious time in trying to prove the existence of something invisible, instead of trying to collectively study and find the origins of things that are visible and waiting to be known. Isn't it foolish then to be engaging in such debates knowing very well that our lives are limited? We can afford to waste most things, but not time. Also, having a debate makes sense if the statement has any scope of being falsified. If the existence of a creator is so strongly real in the minds of the believers, then their eyes will start seeing things that don't exist. Such a debate thus will never reach an end. But humanity will, and that's sad. For thousands of years, we believed the wrong answers and when the right answers arrived, we started asking the wrong questions.

If a deity created the universe, why did the deity wait for billions of years to make man and spread his message?

The reductionist principle of Occam's razor tells us that one must always employ the simplest hypothesis and always follow the common sense principle of simplicity. The large hadron collider just gave us one particle, didn't it? All phenomena in the universe arise from simple causes and theories. It is time we learnt from that. So, which one is a simpler explanation: that human creation happened because of evidence-backed evolution or because of an elaborate creator whom no one has met in thousands of years of faith and worship? I guess it is with our imagination and curiosity that we have put together our scientific understanding of the fabric of the universe and nature of reality. It turns out that the same gift of imagination also had its say in us coming up with the concept of God. The structure and order-obsessed section of my mind has tried to break down the progression of events, as follows:

- Basic life + Evolution = Humans (A)
- A + Time = Intelligence (B)
- B + Curiosity = Imagination (C)
- C + Ignorance = God (D)
- D + Creativity + Fear + Hope = Religion (E)
- E + Time = Religions (F)

- F + Ignorance = Hatred (G)
- G + Fear = Violence (H)
- H + Time = Deaths

What can one make of this? That we are inevitably destined to kill each other in any case?

Destiny means that the future course of events is predetermined, and you are here just going through the motions without any control or influence. This concept arose in ancient Greek philosophy (of course) around the 3rd century BC when philosophers said that all human actions eventually went according to God's divine plan. What do you believe in? That you have free will over your actions or everything is already pre-decided? If this was a YouTube video, I would ask you to let me know in the comments section. (And also subscribe and press the bell icon; never did the humble bell get so much attention in the history of humanity.) A pre-decided life sure sounds cool and opens up a world of the unknown and the magically mysterious. Doesn't it? Let's not get carried away with our fantasies. We have a mind, and every moment is a point in space-time. We decide what we are going to do at any given moment. Simple. The only thing deterring us from complete free will is our social conditioning which influences our thoughts and decision-making to an extent. But with effort, though hard, we can unlearn, become unconditioned and attain wisdom. Thereby leading to complete and thorough free will.

The only other thing that can impact free will is the action of planning in advance. By planning in advance, we tend to alter our actions because factors like stress and anxiety come into the picture. Also, if others are aware of what you are planning to do, then they may impact your action in some way. Thus, actions that are done without any advance planning i.e.

completely impulsively on the fly are the ones done with complete free will. If you disagree and would like to argue saying that it is all an illusion of my mind that I have free will, then all I have to say is that isn't everything that we think is real eventually dependent on our minds? If it has indeed created the illusion of free will, then it can very well create the illusion of any reality and all of us in it. In other words, nothing is actually real. We believe what we believe in, and I, for one, don't believe in destiny. It is a fairy tale phenomenon. It is not my destiny to do anything predestined, and no I was not destined to say this.

Blind belief and conformity devoid of common sense add up to faith.

Man: I have come here to seek help. I am very ill and want to pray to be cured and become healthy.
Wise monk: Illness is a combination of chance & actions. The chance factor is on account of hereditary and genetic defects. Actions include drinking, smoking, and other intentional self-destructive activities. Most times, your actions are a result of your circumstances. Most times, circumstances are beyond your control. So, the chance factor comes in again. Again pointing to the random nature of everything.
Man: I do not understand your divine discourse, holy one. I just wish to be cured.
Wise monk: I am not holy. No one is. Have you been to a hospital?
Man: Yes. The doctor said that the cure is beyond the scope of medical expertise and requires supernatural intervention. He asked me to have faith and believe.
Wise monk: I think you chose the wrong hospital and definitely the wrong doctor. Anyway, I can see that the illness is more in your mind than in your body.

Man: What do you mean by that? Oh, please help me get cured. I want to live and be healthy.
Wise monk: Your mind is blinded by a belief in God and that he can cure you. The belief in a creator and a savior is a drug like no other. It can cure your troubles of reality by blinding you. You then don't see reality so your troubles go away. But the fact is that your mind is now blind. You have lost the power of sight of reality.
Man: But my troubles have not gone away. Also, I can see very well; I am not blind.
Wise monk: Oh, foolish one. You can't look for something with your eyes closed.
Man: I am sorry but I simply cannot understand what you are trying to tell me. Do I need to go on a religious pilgrimage?
Wise monk: You are the victim of a religious upbringing. Among all things that our life works with, the most difficult to unlearn probably has to be religious upbringing, as your life's foundation, your definition of good and bad, and your very understanding of meaning are based on that. It is like an irrepressible parasite that lives in your brain.
Man: No disrespect, but I will go to a different monk to seek blessings.
Wise monk: Good decision. I do not believe in blessings. I am an atheist who shut the door on religion, never to look back.

Your belief that you can fly will not give you wings. Don't you be foolish enough to jump off a cliff believing that you can and are destined to fly.

There is one section of the human population that used to believe in God and now doesn't because of some event that happened in the lives of those humans, like the sudden death of a loved one, or the experience of some other traumatic occurrence. They lose faith on account of disappointment in, and hatred for, the creator. They are unable to cope with the tragedy and blame the creator for it. They stop praying and believing in the almighty. Now, this is another form of atheism but completely for the wrong reasons as you can imagine. The way I see it: humans blame, hate and lose faith in something that never existed in the first place.

The world is full of mostly sane people with insane beliefs.

In her mind, Mother Teresa believed that she was married to Jesus Christ and it was he who asked her to come to India. Maradona's famous goal in the 1986 World Cup was called the 'Hand of God'. Sports teams seek blessings of God before important competitions. When traveling, political leaders of nations visit places of worship. Award-winning people thank God in their acceptance speeches. As a result, we are constantly in the midst of everyone believing in God, especially everyone who is important and looked up to. This constant and convincing messaging, both direct and indirect, is what keeps our minds shut to the actual truth.

If you are praying to God for a good score in your science exam, then either your science teacher has utterly failed in teaching you or society has completely brainwashed you.

Once upon a time, there lived a legendary race of people, the Astomi, who had no need to eat or drink anything at all. They survived by smelling apples and flowers. This is what Roman mythology tells people and foolish people believe at face value. A curious mind should question as to how a human can survive without food and water. But the mind has put curiosity to sleep. It is happy living in ignorance. Such minds need to be shaken up. As a species, humans seek reasons all the time and that stems from our inborn curiosity. It has brought us this far. While seeking reasons is one thing, our arrival at conclusions has been a problem. If most of us are satisfied with the God reasoning for the creation of man and existence, then that reason-seeking quality is not really used correctly. Plus, we stop asking further questions because we feel those are already answered.

Curiosity is always at the door. Some of us just remove the doorbell.

One of the most nonsensical things some religions have done is imposed bans on and mandated changes in some naturally occurring phenomena. Allow me to list out a few:

- If some humans are naturally born homosexual, why does religion ban homosexuality?
- Why does religion consider a life-giving phenomenon like menstrual cycle a taboo and impose restrictions on menstruating women?
- If the natural human form is the finished product of billions of years of evolution, why are religious alterations like circumcision and body piercing done?
- If all the early ancestors of humans ate meat and it played an important role in the evolutionary development of our organs, why do religions impose all kinds of diet restrictions?
- In a world where freedom of expression is important in order to lead fulfilling lives, why are religious restrictions like wearing a burqa imposed?

The burden of proof lies with the person claiming that something exists. This logic seems to have never applied when it comes to the existence of God.

In this society, we are all theists by birth. Some of us turn atheists through the realization of truth. The truth, ladies and gentlemen, is that there is no entity called God. I am beginning to sound like a broken record, but let me say this again. God is a creative solution our ancestors came up with and passed on. We then engulfed our world with fiction, and with concepts that were man-made. But, we can still make amends. We now have a better solution in the form of science, practical reasoning and common sense. We must employ these to answer all questions that cross our minds. We should divert all of our collective attention and energy to gain a true and deeper understanding of the cosmos we live in. But yes, the concept of God was required in an era gone by and we must thank it. In the absence of it, our nomadic ancestors must have gone mad trying to understand all that was happening around them. The institution of religion helped us to an extent in various phases of history and instilled moral values in us so that we could live, think and work together. But now our eyes are open. We are now intellectually civilized and science has answered all questions. Hence, the concept of God needs to be done away with. Thanks to science, we can see the true nature of the world and its origins. Let's not go back to being blind.

The light of reason can illuminate the dark surrendered mind.

In spite of all its flaws, we also need to acknowledge that religion did help in human bonding and the development of society. Men, across all cultures, spent a considerable amount of time in team activities like building religious architecture and studying scriptures. This would have, no doubt, strengthened the bonds between individuals and families, and enabled harmonious co-existence in the society. It also provided an opportunity for men to brainstorm together when solving large-scale civil engineering related problems. There was never a dearth of such bonding opportunities. You only need to take a look at a few examples to understand the scale of construction that's happened in various cultures, which would have obviously required people working together in planning, execution, and maintenance. Such marvelous monuments also provided humanity an opportunity and a platform to exhibit its project management, creative and artistic talents. For example, the Angkor Wat temple in Cambodia is spread across 400 acres, took 35 years to build and required the efforts of more than 300,000 laborers. Five million tonnes of sandstone were quarried from more than 50 km away. I think of the God-worshipping laborers toiling away knowing, at the back of their minds, that the effort was for a noble religious cause and they would be rewarded by the almighty. If the laborers knew that the effort was for a construction project rather than a pursuit in acknowledging and pleasing God, would they have done this with an equal amount of enthusiasm and dedication? I don't know.

An old man, on the verge of death who prays to God to save him, has led a foolish existence all his life.

An addendum thank you note I have for religion is that it has helped us distract our attention from seeking the true grand purpose of existence. We are forever occupied with religious thought, practices, rituals, and contemplation that there is no time to think about why we are here. That's good in a way. I'll tell you why that's the case somewhere in the next chapter but take it from me that that's a good question to avoid, and religion helped us keep our minds distracted. We were just focused on adhering to values, praying for what we wanted, seeking admiration for doing good deeds, and trying to earn a prosperous afterlife and peaceful spirit. What a way to live!

While we are on this subject, let me list down the top three areas where I feel religion will play a critical role in helping humanity scale new heights, as a species, in the future:

1. Your search did not fetch any results.
2. Error 404. Area not found.
3. A system crash has occurred. Universe will restart now.

When you are on the wrong road, how fast you drive does not matter.

The group of atheists (including agnostics) has always been in minority. Some studies say that it is around 7% of the world's population. That makes believers a big majority. Such is the might of the religious institution that it is forever allowed to punish or convert non-believers. I fear that if the world is eventually filled with only pious theists, the part of our brain that's responsible for reasoning will slowly become redundant because of non-usage and eventually be phased out by evolution. Makes sense? We will lose our ability to reason and can say goodbye to any pursuits of attaining wisdom and seeking the grand answers. I do not wish to exist in such a world. I don't want God worshipping men talking to me about fact and fiction.

When every being and every phenomenon is connected to the divine, nothing remains connected to reality.

There is no divine plan that you fit into. In fact, there is no plan of any sort. Things just play out devoid of a plan and yet fall in place. All supernatural things that we believe are real are just in our minds. There is nothing out there in the real world. They are like the delusions of a schizophrenic person. All too real for him, but we know the cause. As I have mentioned before, a child will never ask a question about the supernatural or God until those concepts are taught to him. The fact that such concepts are not ingrained in the natural genetic construct of a child should be enough proof to come to a conclusion on what is real. Evolution and nature do not lie. The vessel of science, and not religion, will take us to the ultimate truth and final enlightenment. Look at it this way: when humanity is faced with an extinction threat via a natural disaster or asteroid impact, who will we summon? Scientists or priests? Well, there lies your answer.

Science takes you on a ride to explain how everything works. Religion takes you for a ride to explain why everything works.

Science and religion locked horns in a famous public debate that took place in Oxford in 1860. The biologist Thomas Huxley and bishop Samuel Wilberforce entered into a heated exchange on whether human creation happened because of the Darwinian evolution theory, which was published a year back then, or it happened because of the accepted theory of God's will. Topics like the human evolution from monkey took center stage as Wilberforce asked Huxley on whether it was the resemblance of his grandparents to monkeys that he deduced his ancestral descent from apes. Huxley replied saying that he was more ashamed to be connected to a human who preached blatant lies than to have monkeys as his ancestors. That must have been a fascinating discussion. However, the absence of a decisive higher jury has long affected the battle between science and religion.

The scientist in me needs to see in order to believe. The priest in me needs to believe in order to see.

When in excruciating pain, you will take any painkiller you can get your hands on. I find this analogy apt for religion and science. Our excruciating pain is our inability to explain the origins and workings of everything around us. The temporary painkiller is God and religion. The permanent cure for the pain is science. The painkiller helped us deal with the pain for a long time, but overuse carried its side effects. Thankfully, science has come along as the treatment aimed at the root cause. We should welcome and adopt it, and thank religion for its contribution. I repeat that it is now time for religion and the concept of God to go. Some niggles like the questions of the creation of the universe still remain unanswered. We should let philosophy hold forte on that subject until science finds the answer.

Belief in anything without evidence is plain wrong. It undoes millions of years of evolution that got us our intelligent brains. We are letting nature down by not living up to our intelligence.

Patient: I wonder why God made my teeth crooked.
Wise dentist: I doubt if he had a hand in that. You see, the human jaw size has reduced with time as we stopped using our teeth to dig into and tear apart meat once we invented tools. That being the case, a smaller jaw has thus resulted in the teeth of some people being crooked.

Patient: No. I believe that everything happens as per the grand design of God, including our meeting at this point and me having my crooked teeth corrected. God endowed me with the wealth to pay for this treatment.
Wise dentist: I see that you are more religious than many of my patients. I should start keeping a copy of some religious scriptures in the waiting area, alongside all the comics.
Patient: God has endowed you with your skills and you lead a happy life, thanks to him.
Wise dentist: Quite the opposite. My wife has filed for a divorce and I am knee-deep in business losses. Plus, I don't believe in God. I feel that extreme teachings of religion kill the intellect required to ponder over profound questions about creation and reality.
Patient: You must pray to the almighty to get you out of your troubles and open your eyes.
Wise dentist: I am afraid that God hasn't given me the medical skills to fix crooked minds.

Evidence speaks for itself. And the lack of evidence, for thousands of years, puts it in writing.

If you are a devout follower of God and have been so all your life, it is highly unlikely that this chapter would have had even an iota of impact on your beliefs. Denial is always our immediate and preferred response as humans. Our whole foundation of understanding is based on religious beliefs and without them, we wouldn't know where to start and what to do. It is a way of living for nearly all humans and comes as naturally as breathing. It is so real in our minds that we see it as actual reality. I understand all that. Surrendering yourself to a supposed superior authority, to take care of you, is always the easy way out, and that's what keeps religion going. Humans are so dependent on the religious system that they are incapable of surviving without it and hence will go to any lengths to defend it.

Those who surrender themselves to the illusions of a creator, faith, and destiny give up their ability to swim in the great ocean of truth.

If questions about existence and the pursuit of the absolute truth are not on your radar, then you can happily carry on with your content and blissful life of prayer, morals and pious deeds. In simpler words, you can continue living a lie. But if you are interested in understanding the true meaning of the cosmos, then it would be advisable to make a U-turn as you are driving in the wrong direction. In this context, I can think of a famous quote by the British statesman Winston Churchill to wrap things up: "Men occasionally stumble over the truth, but most of them pick themselves up and hurry off as if nothing had happened."

PURPOSE & METHOD

I went to the woods because I wished to live deliberately, to front only the essential facts of life, and see if I could not learn what it had to teach, and not, when I came to die, discover that I had not lived.

– Henry David Thoreau
Walden (1854)

We are fragile tiny beings that are staring at the infinite universe and trying to understand what exactly is the purpose of our lives? For what reason do we exist? Why are we here, and what are we supposed to do with this gift of life? How do we go about the business of living, what do we aim for, and how do we motivate ourselves to get there? Do we need to construct purpose or is it something already existing and we just need to find it? Is my purpose different from yours and if so, why? Will the purpose get us somewhere or will it forever be a pursuit? Will there be a guide to help us or all of us need to walk the path alone? Are there clues hidden along the way to aid us? Is it a fun ride or a painful drag like this book? Is it optional, in the sense that can I skip it and laze around all my life instead? Whose purpose was it to come up with this idea of a purpose in the first place? Wow, that's a lot of deep questions. But worry you must not, for this chapter will guide you to the deeper answers. At last, we are finally here. After having examined, at somewhat painful length in previous chapters, all subjects that would cross the mind of a curious being, we are ready to discuss the most relevant matters. Praise the Lord, shall we then?

I know that you appreciate the fact that I have come across as a straight shooter so far and I won't dilly-dally here as well. It seems that the greatest trick life ever pulled was to make you believe that there was a purpose to it. Yes, I have realized and you should be ready to face the truth that life has no purpose in the grand scheme of things. In the immense and unfathomable size and time scale of the cosmos, all that us earthlings have done, are doing and will do, is inconsequential. It is a small drop in an infinite ocean. There is no contribution that the cosmos seeks from us. Neither does it value should we try to make one, for it is forever indifferent, as we have come to realize. Evolution, through random mutations and natural selection, created us but at no point ever did it have any purpose in mind. It just completed a task, devoid of any intent and emotion. If our creators did not have a purpose in creating us, how can we have a purpose while living out our existence?

For what purpose does the grand cosmos exist? None. It does not exist so that it can create galaxies and solar systems where planets like ours can harbor life and create organisms that can acknowledge the cosmos' existence. In the very first chapter, we had discussed that life has to exist in the universe because, without it, the existence of the universe would have no meaning. But did the cosmos create life for that purpose? I don't think so. Life happened, like how galaxies and stars happened. All of it happened completely randomly without a plan or design. The physical laws exerted their forces on matter and energy, and celestial events took shape. All of it transpired without any predefined purpose. Also, the universe does not have a mind or consciousness to even regard the concept of a purpose. The cosmos is purposeless, and so are we.

For what purpose did evolution happen? The same logic applies here as well. We know that evolution is possible only because genes do not replicate perfectly and thus lead to random mutations. This basic property just played out on our planet and all species came alive in this Eden. On other planets, if there are organisms with genes that replicate imperfectly, then the process of evolution will come into effect there as well and start natural selection forever. If those organisms are different, then evolution will not take place on that planet. That's that. Evolution is just a simple process, with mind-boggling results like giving rise to us intelligent humans. But it is just a process that occurs naturally, devoid of any plan, purpose, joys or regrets.

Executioner: You are to be hanged tomorrow. What thoughts occupy your mind on the eve of your death?
Convict: I have no thoughts. Everything was pointless to begin with. Your, my and everyone else's existence has no purpose. Hence, I have nothing to remorse or rejoice at the onset of death.
Executioner: You seem to have entered depression.
Convict: The room of truth is dark. We don't know where the switch is, or if there is a switch at all.
Executioner: The room may be dark, but we can help each other in finding our way.
Convict: In the absence of the great truth, we're all blind. A blind man cannot show the other blind man the way.

Executioner: Thousands of intellectual men have worked for centuries to make discoveries that will eventually lead us to the right answers.
Convict: Who is intellectually fulfilled? How does one become so? In the absence of truth, only the lesser minds feel that they are fulfilled. In the absence of knowing the real truth, all conclusions they have arrived at are just premature and incorrect closures.
Executioner: Why do you inflict so much pain on yourself? Realize that ignorance is bliss. When on board a cruise, do you want to enjoy the sights and perks, or do you want to spend the journey figuring out how and why the ship was built?
Convict: The unbearable suffering of my existence of struggle can be reduced only by realizing the purposelessness of it and everything else. Everything is futile. But the futility of it is not so straightforward.
Executioner: You are not at all taking into consideration how we have reached this far from our primitive beginnings. The human spirit involved in getting us here needs to be celebrated.
Convict: We will lead out our lives on this planet, as humanity makes some inconsequential inroads into understanding the real truth of the cosmos. I agree that the human spirit always refuses to give in. But the pursuit is in vain, so there is no point romancing the human spirit. In an exercise of futility, being wise or a fool makes no difference.
Executioner: The problem is that we are in that time and place where we cannot prove anything with absolute certainty. We are presently incapable of understanding a lot of things. Let's be humble about that and make peace with it. The great truth of reality is definitely beyond religion, but may also be beyond science. However, scientific research via empirical evidence must be the road that should be taken.

Convict: I am a little wary of scientific research. You see, I am often observing things from the point of view of others. Then, is that true research for me? We somehow refuse to accept the fact that we live in an unimportant point, both in space and time.

Executioner: Those who don't question the why and how of existence lead a delusional life. Know that you have lived a good life for you have pondered over the puzzle of your existence.

Convict: Yes, but for what? The only tools we have to find the ultimate truth are our senses and our brains. Not a lot to work with.

Executioner: I can empathize with your state of mind, but realize that the point is that you're not here to prove a point. You are just here to experience existence. Find ways to experience it happily. Don't keep searching for witnesses for events that probably did not take place.

Convict: In that case, the God hypothesis is so much simpler. It explains everything, teaches morals and goodness, and ensures a life of peace and happiness. If all existence has no purpose and we are just meant to pass our time here, we should accept the God hypothesis and just be. But that would be so wrong and a life of blatant lies.

Executioner: While I agree with that, riddle me this: If all pursue the path of truth, then who will take up roles required for a civilization's existence? Like cooks, postmen, doctors, laborers, etc. How will society function?

Convict: Sometimes I feel that the laborers have figured out the purposelessness of it all. They are happy all of the time, with or without wealth or achievements. They seem to have cracked the golden motto, which is to do brainless work all day and chill out in the evenings.

Executioner: I see. Is there any last wish of yours that I can attempt to fulfill?

Convict: I choose to break my head, to feel disoriented, to feel suicidal in solitude, to cry in the misery of being forever ignorant. Hand me the truth. Set me free.
Executioner: That I cannot do. Also, this conversation is not headed anywhere. I suggest you call it an early night. You have a big day tomorrow.
Convict: My consciousness arose from nothingness the day I was born, and I am looking forward to returning to that state of nothingness tomorrow.

In his book 'The Tao of Philosophy', Alan Watts talks about how things like dance are purposeless. He says that the aim of dancing is to just dance, and it has nothing to do with arriving at a particular place on the dance floor. He talks about how we do things like going on a walk with no particular purpose in mind, and it is that purposelessness that makes us rational human beings. I couldn't agree with him more. The Sun does not rise every day because of a purpose. It does that purely because that's how its existence is. The asteroid that hit the Earth to wipe out the dinosaurs did not have that devious purpose in mind. It had no purpose in mind. Things just exist in the cosmos, devoid of purpose. Granted that things like asteroids and the Sun may not be conscious beings, but consciousness only gives us the power to make sense of what's happening and it does not give anyone any purpose for existing. It, in fact, helps us realize that there is indeed no purpose.

So, does that make us robots that are just going through the motions? In some ways, yes, and many ways, no. We are conscious beings and hence consciously aware that there is something like purpose that things can have while existing. Robots don't have that realization. But, we are like robots in the sense that we carry out actions on the basis of thoughts that are programmed in us via our genes, upbringing, experiences, social conditioning and a tad of free will. Thus, in a way, we go through the motions like robots on the basis of that programmed information. Man, that's depressing! But hold on, right there. There is a lot of joy and satisfaction to be experienced while going through the motions. There is serene beauty in purposelessness should we choose to focus on the right things.

Let's look at it another way. I had no say in me coming into this world. I did not ace an exam or win a lottery to get a chance to be born. Nor was I born with any tremendous sense of accomplishment and pride thinking that I had defeated 249,999,999 other sperms to win the race. I was brought into this world without my will or efforts. After making the coveted entry, I had no say in how I was raised and taught. I had no say in how tall I became and the color my skin took. I had no say in choosing the society I was raised in and which shaped my thoughts. So, in all of those, I have just been the recipient of the effects of causes not caused by me. How then can I now turn around and say that I have a lifelong willful purpose that I am forever focused on? Wouldn't it be so wrong?

You can choose to live your life thinking that it is full of purposes, on a daily or a longer basis. Your immediate one is to somehow finish reading this damn book. Then there's something long pending that you need to finish soon. Also, there's something else that needs to be done by the end of this month perhaps. So, there are so many little purposes that you are working towards and will derive pleasure upon completing. Well, these are not purposes at all! These are action items in a checklist. Why you made that checklist is related to the concept of your purpose. The eventual objective of living your life is what purpose stands for. In other words, purpose is the ultimate reason for which you exist. That, needless to say, cannot be part of any checklist. People create bucket lists and think of it as the purpose and aim before death. Wrong again. I guess the first purpose in life would be to understand the meaning of the word purpose.

What's that? You're saying that your purpose is to lead a peaceful life as a kind human being doing good deeds and making the society better. What does that sentence even mean? The human society is on its way to extinction and even if we figure out a way to colonize other planets, those will also die one day. Doing good deeds is a way of living and not the purpose of living. Being kind is just a choice of conduct that you would like to have. You may forever live a peaceful life in your mind as a kind human being doing good deeds and thinking that you are taking the society forward. Life will go on. You will continue being yourself and perhaps may change a little, from time to time. You will have periods of happiness and times of sorrow. That's pretty much it and then you will die one day feeling content in your mind. But once you are gone, those alive may remember you as a good person for a while but you will eventually be forgotten. How useful were any of your actions then, I ask? Besides, who was it supposed to be useful for in the first place?

Man: Good lord, how many ants are there in this house!
Wise ant: Why are you so bothered by me? I bring no harm. I occupy a small space in your house and go about my business without causing you any trouble.
Man: You should go live outdoors where you belong. You impact the cleanliness of my house. This is my space.
Wise ant: I am like the Earth in the cosmos. Does the Earth make the cosmos dirty?
Man: What the hell do you mean by that?
Wise ant: I appear in some remote corner of your house for a little while and then disappear. What significance do I have in

your life? Nothing, right? The Earth is like me. Its brief appearance does not matter in the universe's life.
Man: Oh, wow! Was the Buddha your teacher?
Wise ant: The Buddha had no teacher.
Man: So, your analogy then makes my house the universe?
Wise ant: Is that all you understood from my analogy? I am glad that I am not a part of your foolish and cruel species.

Facebook has close to 2.5 billion users worldwide. If everyone on this planet had a life of purpose, social networking platforms would be not so prevalent and popular. Don't you observe how we, unintentionally as it may be, try to keep ourselves occupied and distracted so that our weaknesses cannot dawn on us? In order to create a life of fulfillment, we need to think a lot and tax our minds all the time. We need to use our intellectual power to ponder over questions of all kinds, including profound mysteries. But people are unwilling to think for some reason. They have been conditioned not to use their minds that way. Social acceptance, conformance and earning a livelihood are all that matter. We are afraid that we might discover that we are intellectually inferior should we give it a shot and come up short. So, we distract ourselves by engaging in everyday trivial matters and pointless conversations, hoping that we will somehow find fulfillment and purpose, or won't be bothered if we don't find them. As a result, we become habituated to following the crowd. We run away from things to look for those very things. This is how the psyche of a majority of our species works. A major chunk of humanity lives in complete oblivion. They don't know how and why they are here, and what they are supposed to do. The meaning of the word purpose is not understood at all, and then even the wrong meaning is avoided. We are creatures who refuse the power of sight and want to intentionally remain blissfully blind.

This apparent purposelessness of existence seems to get a foot in the door when half of your life is over, and then slowly keeps making its way through all of your remaining life. It grows on you as you grow older. I like to think of it as one of the aspects of wisdom. With time and experience, you gain wisdom in life and then the pointlessness starts hitting you. By the way, the funny thing about wisdom is that it comes when most of the important decisions in life have already been made. So, you can only use it to mostly unlearn and do some amount of damage control. Pity. A wise man once shared with me the irony that wisdom comes once you reach the point of no return. The knowledge that he was happiest when he stayed single came after he had children. The knowledge that his calling was in traveling the world came after he spent all his savings to buy a house.

Man: I have come to become wise. Please teach me wisdom.
Wise sage: In the great schools of the intellectual world, there is no course on wisdom. My friend, it cannot be taught. I can only share my experiences. You must then spend time and energy to experience, think, envisage and attain wisdom.
Man: But that will take a very long time and I may fail. What good is your wisdom if you don't share it? It will not diminish by sharing.
Wise sage: The wise do not take desperate measures. Be patient and entertain the right thoughts. Wisdom will come to you.
Man: Are there any guidebooks that can help expedite the process?
Wise sage: You can read a book every week. But unless you think and contemplate, it is of no use.
Man: I will try. How will I know when I have become wise?
Wise sage: When you refrain from using your intelligence every time and everywhere.

Wisdom is such a wonderful thing. It can make you see things so much clearer. It can expedite the accumulation of rich thoughts. We had learnt in the chapter on evolution about how being dumb helps you survive. Perhaps that's the reason why nature has made wisdom so elusive as it can lead to things like the discovery of life's purposelessness and make some disappointed people plunge into depression, thereby risking their survival. Living in delusion helps you live longer, apparently. What good is it though? Nobody is interested in playing a video game that's long but increasingly boring.

If I were the reader of this book and had not ever wondered about the aimless nature of my existence, I would obviously say that this book talks rubbish. I would say that clearly, my life has meaning. The very reason I choose to not end my life and find a reason to wake up every morning are clear indicators that I have a purpose. But I would be a fool by thinking that way. I wake up every morning because I have taken on some responsibilities towards the ecosystem I have built around me. This ecosystem consists of my family, society, and my work. There is a role I play in that and that keeps my mind and body occupied all the time. Yes, I agree that there are responsibilities I take on if I start a business or become a parent. But, responsibilities do not add up to a purpose. Rather, a purpose breaks down into responsibilities.

Well, if there is no purpose then what's the point of existing, you may ask. Do the minds that already know about this, in all probability, suffer from collective depression? Owing to the notion that true reality & purpose of existence are not very uplifting. I encourage you to think of us as visitors in a museum. Our time is to be spent in understanding all that we can experience and hence make good use of our time there. We should try to understand everything deeply, come up with the right questions and look for answers. The museum has no tour guides and things need to be figured out on our own, and satisfaction gained by doing that. This pursuit of ours to find the great truth is what our purpose is in our 'local' scheme of things. The absolute truth may be impossible to find, but it is also impossible not to look for. Our role is to understand the grand cosmos we live in and everything related to that. The ecosystem you are in should enable you to pursue this purpose, and you should guide your ecosystem towards this reality. To reiterate, our lives have no discernible purpose in the grand scheme of things but we have lots to do to make proper use of our limited time here. So, purpose exists in the smaller scheme of things. Unless, of course, Earth is the only place with life in the cosmos, and thus all that we understand about the cosmos is the only way for the cosmos to know itself, which then makes us all special. But, that's pure fantasy in my view. We all always want to feel unique and special, and that's not part of the purpose. Our greatest achievements as a species will always be the steps that got us closer to understanding the universe and our place in it. Nothing else is relevant. Surely, Thoreau said it right when he proclaimed, "Rather than love, than money, than fame, give me truth."

But some of us didn't arrive at this conclusion easily, and a majority of us do not know and will never believe that we don't have a purpose in the larger scheme of things. This holds especially true for the religious among us. Men have been looking for a purpose of existence ever since they developed some basic intelligence. While our ancestors philosophized on this question, they also had to deal with lurking predators and famished bodies. We don't have those problems to deal with and hence have been in a much more comfortable position to find the answer to this question. But as it turns out, it is this comfortable position that has come in the way of us contemplating the purpose of our existence. How? Well, we are so caught up with the evident mediocrity of our everyday comfortable existence attending to trivial matters, that there is no time and mind space for us to think of such higher matters. If we rid ourselves of the little comforts of engaging television shows, movies, sports, foods, societal wins, and everyday routine, then the mediocrity of human existence will eventually dawn on us. The effort to find purpose begins then.

Thanks to our upbringing, our conditioning and everything else that our lives revolve around, the lives of many have only three parts:

- birth
- efforts to get accepted & appreciated
- death

We live our lives just trying to accomplish material things, make our name, feel and be considered successful, feel good and devoid of any regrets. And then we die. What good is such a life? What good did we do with the miracle of life that we were blessed with? Did we try and understand the bigger miracle of the existence of the cosmos and the origins of things like time? No amount of wealth or assets will help you lead an existence that is anything other than plain vanilla ordinary. It's only when you comprehend and accept the bigger picture and reality does your life go beyond being ordinary.

The purpose is not to spend time trying to look for a purpose. If you exist just to extend your existence, then you need to take a moment and think. There's no doubt that answering profound questions about creation is not required in order to survive. But, it is not necessity that drives us. It is our curiosity to know, and that's a sign of an advanced civilization looking for a purpose, isn't it? However, are we all on board this journey? There is perhaps a reason, for no particular reason, why there is so much to know and understand about the world around us. Being content with not knowing is not in our basic construct. The quest to know drives us as a species. In a universe with all the answers and everything figured out, we would have no drive to exist. I am not trying to come off as an enlightened sage here and trying to preach. I have seen too many people being occupied with things that don't matter and they're completely oblivious of it. When some of that oblivion tries to fade away, they are not willing to face the truth. Think of it this way: you have not been given time on this planet to kill time, to preoccupy yourself with trivial things that don't matter, and to accumulate things that have no real need. So, don't kill time. Don't run away from boredom. Face your thoughts. Gaze, think and wonder. Find meaning in your life. Then you won't have enough time on your hands to afford to kill any of it. Purpose may be the quintessential human delusion, but it is also our greatest driver.

Okay, I don't want to sound like a broken record and thus assume that you are willing to momentarily divert your focus from the immediate daily deliverables and are ready to ponder over some cosmological mysteries. It's okay to skip a conversation about the weather and sports in order to make some time. What kind of mysteries, you ask? The quest to come up with a unified science theory that works on all scales, to gain a better understanding of dark matter & dark energy, and to delve into mind-boggling phenomena like black holes—these are some of the summits yet to be scaled in our expedition to find the great truth of the unknown. Or, we can join the pursuits to understand how the mind works, how alien civilizations can find us and how time can have a beginning. Or, we could possibly think about whether there are other universes like ours, and what created all of them. Great. Will do for you! What next? Well, there's no next. The purpose is just that. To dwell over, keep looking for evidence and basking in the glory of such questions with the hope of answering them correctly. Several scientists, philosophers, and the like have realized that without this contemplation, they have only the company of vacuity that engulfs them always. We better listen to them. In the process, we can hope to accumulate wisdom, which is something you've read about one too many times in this book I am sure. Our curiosity for the ultimate truth will help us eliminate the mediocrity and embark on a journey towards having a meaningful existence.

What's that? You are saying that such questions have answers that can never be found and thus what good is any time spent on them? Well, I agree that we will probably never fully understand the workings of the universe, but we should never ever give up the joy of the effort and small milestones. You will not lead a wasted life for sure. The journey itself is so fulfilling that it makes up for a possible non-arrival at the intended destination. And we become wiser along the way. That's an offer that cannot be ignored. Plus we get to have a crack at some answers, and being wrong is absolutely fine. So, why fear and resist? The beauty of science is that it is okay to be wrong. You are encouraged to think out loud with complete freedom. No ideas are suppressed. Your theories undergo the test of data, observations, and critique. You may end up being right or wrong. Doesn't really matter. What matters is the continuous pursuit of seeking truth and understanding, and if you come across something so complex that you can't even understand it, let alone attempt answering it, then that's perfectly alright. The simplest solution for the most complex puzzle is to not solve it, for it cannot be solved. Just understand it fully and appreciate the beauty of its complexity. Deep meditation on such puzzles can lead us to some sudden and striking realizations.

Having said all of this, the easiest way is to go believe in God, pray every day, feel that you are spreading goodness and will eventually be rewarded. Eat comfort food, watch fiction on smart screens and feel good. Look forward to little pleasures like upcoming travel and social events from time to time. And that's that. You'll live your life without much discomfort. Sure you'll have some tough times, but no big deal. Life will sail through. Would you like to sign up for such an existence? I pray that you don't.

Cow: Observe me carefully. My life is focused on meeting survival needs and once those are met, I am content. I rest on the land in complete peace, being thoroughly fulfilled. I am not focused on studying my origins, worrying about global warming or colonizing other planets. I am just here, in the moment, content and at peace. Why are you so unhappy and restless, human? Why don't you be like me?
Wise herder: Sometimes, ignorance is bliss. Not knowing will keep you happy and content, whereby rarely will questions like the meaning of existence cross your mind or bother you.
Cow: Exactly my point. If life is short and pointless, why don't you just be happy?
Wise herder: The ride of life is limited in terms of time but unlimited in terms of mystery, wonder, and knowledge. I have made science my friend. I spend time with it to understand reality, travel in my mind to different worlds, ask profound questions, and search for answers. I choose to let every moment in my life be a glorious chapter in the everlasting journey of exploration. If in that I am briefly unhappy from time to time on account of all the difficult puzzles involved, I am happy with that.
Cow: You do have a way with words. I am sold. How can I make science my friend?

Wise herder: It's never too early to join the pursuit of science, and never too late. All species are welcome, of course.

Many life coaches say that if you don't wake up every day excited and energetic to get started with work, you are in the wrong profession. I think that applies to life as well. We shouldn't rest in order to fall asleep. Rather, we should rest in order to wake up. If we embrace a life of truth and join the quest of understanding everything around us, then we won't need an alarm to wake up every morning. A mind that researches deep questions about the why & how of creation does not have a single second to waste, let alone feel dull. Your natural curiosity is your ticket to travel to places you haven't been to before. Keep this ticket safely and be ready for an adventure every day. The company of like-minded people you find in your travels will only make you wiser.

Before this chapter turns into a collection of my motivational quotes, I'll pain you with one final reference to HR (human resources), in line with the pattern that's evolved through all the other chapters. In an organization, the HR leader works with the executive leadership and board in coming up with the mission statement of the organization. The mission statement is like a purpose statement, defining what the organization's eventual aim is. For example, the shoe company Nike's mission is to bring inspiration and innovation to every athlete. So, that is the purpose of Nike's existence in the cosmos. What happens if the athletes no more find purpose in running or turn to natural methods after understanding nature's recommended running footwear through tribes like the Tarahumara in Mexico? In simpler words, for an organization to fulfill its purpose or reach its mission, it must bank on all components in its overall machinery to fulfill their individual purposes. This is not easy. An employee who has seen the light of curiosity and who is stuck in a dull job will lead a life of unhappiness as his mind wanders towards exploring the unknown. If he takes the plunge and leaves the job to explore the cosmic abyss, his life will still be largely sad as creation answers are hard to find and fathom, and the world will not agree with him. But his existence will have some meaning and purpose.

Society has conditioned us to think that the purpose of life is to have a successful career. Find what you are good at and excel in that field. I have some questions on this and believe that our education system, by centering its focus on marks and ranks, has ensured that no one ends up educated. What exactly do you mean by a career? How do I build a career? How can I be interested in only one thing and spend all my life in that field alone? Seems like quite a foolish way of leading a life. I am interested in a lot of things right now, and many more in the future. I want to explore all of them, experience by doing and researching. I wish to live the natural way and do what makes sense to me. I have not enrolled for a race, and am not competing with anyone. I have nothing to prove, nobody to impress and am not seeking any recognition or appreciation. I want to be so immersed in my different areas of interest that such thoughts don't even cross my mind. And yes, I am not here to lead a so-called 'successful' life or become an expert or authority on any subject. My life is not about being in the good books of anyone and everyone. I have my own book to write.

During the Renaissance period in Europe, the definition of a well-rounded gentleman was that of an amateur who could write, sing, dance, compose, paint, philosophize et all. Specialization in one field alone was not considered to be of much use. Isaac Newton was a mathematician, astronomer, physicist, theologian, and author. Benjamin Franklin trained as a printer but also became an inventor, scientist, author, postmaster, humorist, civic activist, politician, and diplomat. Even if we look way back around 300 BC, the philosopher Aristotle wrote on subjects that covered physics, biology, metaphysics, logic, ethics, theatre, poetry, music, psychology, linguistics, politics, and economics. Possibly, these men gathered so much diverse knowledge in order to not turn any knowledge seeker away saying, "Not my department." It is very evident that true wisdom comes when you expose yourself to all areas. Being a subject matter expert on one subject alone will get you respect, but not a well-rounded perspective of the world. If you are starting from scratch midway in your life and really want to understand the sum total of all knowledge, I would urge you not to try to become a scholar in only one field. Become a dilettante in multiple fields and pursue, self-study and understand various different subjects. You'll get the perspective that helps in putting all the pieces of the puzzle together.

Most lives have four parts:

- Struggle to exist
- Exist well
- Question existence
- Cease to exist

Some lives start and end at the first bullet point. It's sad to note that not everyone in society has a chance and opportunity to become educated and get exposed to the mysteries of the universe. This is because of the ill-constructed class-based society that has taken shape. A section of the society toils away in poverty, with the sole daily purpose of making ends meet. But, those who are privileged enough to be on the pursuits of wisdom can help the underprivileged through economic means as well as by sharing knowledge. Remember that one day, you will die and nothing that you studied, thought of or wished to do, will ever matter. Your only contribution that would matter is how you shared your knowledge, insights, and wisdom through teaching, writing, or other comprehensible means.

The 'Feynman Lectures on Physics' is possibly the most popular physics book series that has ever been written. The books are based on lectures given by Richard Feynman, an American theoretical physicist and Nobel laureate, to Caltech students during the 1960s. These are considered to be one of the best and most comprehensive introductions to physics for students, owing to their simplicity and the enthusiasm and great clarity with which they are presented. But, not everyone was in Caltech in the 1960s and not everyone in the world has wealth to buy books. So, how does humanity get access to and benefit from these great resources of knowledge? Well, in 2013, Caltech teamed up with the Feynman Lectures website and made the books freely available on their website. That's a great thing they did. So now, with just an Internet connection, people have access to free and world-class education. There's no doubt that intellectual wealth stored safely in the vaults of the brain is useless. It needs to be shared. In that context, this book can be considered to be my small contribution towards that cause. My road to salvation, in a way.

No road is easy. Plans don't go as per plan. When making a foolproof plan, one should always plan for the most likely eventuality that nothing will go as per plan. Our thoughts are not always on our side and largely never in our control. As true as that may be, our thoughts are all we have to find and process the truth. We're not going to attain wisdom through powerful shoulders or chiseled abs. An untrained conditioned mind will always drift away towards negative and mediocre thoughts. Does it then help to think of our thoughts as a different entity? Should you distance yourself from your thoughts and see them as another being? Then, we can befriend this other being. Why not? Let's learn to reason with it. Fight with it. Learn from it. Make it our companion on our quest for truth. We need to train our minds, including our subconscious, to think in a specific way, which is aligned towards the ultimate goal. Your mind is pregnant with the potential for great intellectual and sophisticated thoughts on the most profound matters. Leave the mediocrity aside and allow it to give birth.

We are social animals living in a society. The mindsets, behaviors, and lifestyles of those around us are bound to rub off on ours, however hard we may try for them not to. You will find some very happy people around you who carry sound reasoning on why they are that way. They are those who are not bothered by or turn a blind eye to questions of existentialism or the meaning of life, as they believe that there is no correct answer to that question, and not knowing the answer will certainly not keep anyone happy. Their argument is that if a lifetime is spent better being happy, then why chase the question of existentialism at all. They reason that do we really need to know the answer, and are we prepared to know it at all? You should pat such people on their shoulders, tell them that the grapes are not sour and the perceived happiness of their intentional ignorant existence will eventually bite them at a later point in life. They need to know that if they don't feel any vacuity from time to time, they are on the wrong road. The vacuity adds up and presents itself in one overpowering shot, and that can never be a good moment to be in. In the continuing absence of the great truth, all of our lives are a lie and nobody wants to live a life full of lies.

The lives of organisms of some other species give us clues on the method of living. Dolphins, with their large brains and supposed high intelligence, seem to have figured it all out. All they do all their lives is play happily in the shallow waters. As nature has not equipped them with opposable thumbs or advanced bodies that can do cosmology or any scientific research, they seem to have realized that as long as sustenance needs are met, there is no point complaining about anything. Perhaps in their minds, they philosophize about the origins of all the water they forever swim in. In the nights, they look up in the sky and connect the star dots together in ways we'll never know. We'll also never know what they talk to each other about. They must be so much smarter than us struggling class.

I have always wondered about what motivates other species to exist. What do stray dogs on the side of roads (in third world countries whose purpose for existing is to remain third world) ponder over when lazing around? I guess species with requisite intelligence do not have the mind space to think of such matters. Their existence revolves around finding food and fending off predators, in line with nature's plot. They cannot afford to take their minds off those important matters, to be able to look at the world as an artistic and philosophical masterpiece.

Coming back to discuss more about the supposedly most intelligent species, many of us are mistaken that getting rich is a purpose in life and rich people tend to then find a higher purpose as all options become affordable. You see, richness comes at the cost of time and mental health, both of which are essential for seeking wisdom. Also, what exactly is the definition of richness? At what bank balance figure do you become rich? There is no exact answer to that question. Wealth brings peer pressure, social camaraderie, enemies and stress. Let me tell you about Ranjan Das, who, at the age of 42, became the youngest CEO of the Indian arm of a large global organization. He was at the pinnacle of success and wealth. His job was highly stressful, and he used to resort to running marathons in order to de-stress. He was a very ambitious person, focused on scaling new heights in the corporate ladder all the time. His ambitions, workload, and lifestyle meant that he could manage only 4 to 5 hours of sleep every day. This took a toll and he suffered a massive heart attack and passed away at the age of 42. His death is a reminder that material success is a never-ending game that makes one lose focus on what's important in life. At the same time, some people are able to manage it and live long healthy lives. I guess the secret lies in balance. If you can get rich and still retain your curiosity, sanity, and seeds of wisdom, then that's a way that works and won't kill you. You can then leave the race when you wish and do things that bring wisdom on the road of truth.

We entered the world empty-handed and will exit that way. It only makes sense then that we always lead a minimalist lifestyle. Away from all boasting and displays of wealth, away from unnecessary clutter, and definitely away from coming across as richer than your neighbor. One should realize that when you lead a minimalist lifestyle, you are never rich and never poor. I see poor but happy people in the streets and think to myself that anyone who is poor but happy has figured out the pointlessness of accumulating wealth at the cost of losing out on wisdom. Anyone who is rich but sad also has. That's sad.

Teenager: I have an interview next week. If I can do well, I can grab a comfy job as a Manager in an MNC bank. That would set me up for a comfortable life.

Wise career counselor: You clear an exam because you don't want to fail. Why sit through in the first place?

Teenager: I'm sorry. Didn't get that.

Wise career counselor: Do you want to spend your life going through the motions in a predefined template, or do you want to strive to unearth the ultimate truth?

Teenager: I want to live a happy life.

Wise career counselor: What's your definition of happiness?
Teenager: Well, a well-paying job, marriage, child, house, car, and a good bank balance. I also want to become the best in my profession.
Wise career counselor: You have just listed down a set of desires that will completely consume your time, peace, and life. Also, if success is defined as superiority to other beings, then is it really worth the effort?
Teenager: I know that you are giving me advice on morals and everything, but I just want to get somewhere in life and get out of the entire struggle I am in right now.
Wise career counselor: I, like many, used to lounge in the comfort of ignorance. Now, I struggle in the toil of knowledge. But I am happier.
Teenager: But did you make any money following all that you are asking me to do? I see that you arrived here in a cycle.
Wise career counselor: Always measure one's richness by the things one chooses not to buy.

By now, I am sure that you must be thoroughly fed up with the overuse of terms like wisdom. But I cannot help it. Our preoccupied multitasking minds, with their short attention spans, need to be fed repetitive inputs so that things get registered. Like how songs wishing to become catchy use repetitive phrases over and over. On a serious note, the theme of this book requires reference to those terms time and again, and thus you will keep seeing them. You will also keep seeing the reference to science as the method for obtaining wisdom and contemplating the ultimate truth of the cosmos. Many people make the mistake of understanding science as a huge blackboard filled with physics equations with unknown symbols that they will never understand, and hence oppose science as the path to truth. We should all know and understand that science refers to everything related to common sense, rational thinking, reason-based approach, evidence-based conclusions, well-researched assumptions, prejudice-free inferences, child-like imagination, and the like. Science is the path a curious child would take when not told about anything.

Meditation and philosophy have also helped us realize truths, but those are not the only paths we should be pursuing. In fact, the day when all mental subjective states can be reduced to physical processes, that would be the end of philosophy as we know it. That day may not come soon, but we need to get out there and explore in any case. Let's look in the high mountains, deep waters, and across vast deserts. The theory of evolution by natural selection that explained human creation came about when men wandered into unknown lands searching for answers. Philosophers sitting indoors and contemplating existence could not come up with that theory even after hundreds of years of deep philosophical efforts. You need to get out if you are looking to understand the world. The sudden realization of a great truth is like a shooting star. You'll miss it if you are comfortably indoors.

In the chapter on evolution, we learnt about Charles Darwin who opened our eyes to the miracle of evolution by natural selection. His discovery of the unifying theory of the life sciences came after a 5-year voyage on the ship HMS Beagle, which started in 1831. He spent 18 months at sea traveling, and more than 3 years roaming the lands searching, as it later turned out, for the true creator of all beings. When he joined that voyage as a 22-year-old budding naturalist, his plan was to just visit tropical lands and then return to join the church. The 5-year voyage took him to the far corners of the world, and when he returned, he was a completely changed man, having seen the real truth of reality and met the designer of life on this planet.

We were born like how some frogs are born in a well. But we chose to not live a life within the confines of it. We figured out a way and jumped out and have been trying to understand and change the world ever since. Alongside trying to change the world, we must make an effort to continue to understand it. Only studying existing data will not take us forward in our search for understanding. Our imagination is equally important, and it is the wonder of the unknown that drives our imagination. Our imagination gives us a window to see other places in time and space. It has also helped with so many tools to be at our disposal to aid us in our pursuits. That's because great inventions begin with an act of imagination. From powerful telescopes in space to intelligent computers on land, the scientific community has invented a whole lot of fascinating equipment for use by all those who are willing to. If you are in a career and environment where you are equipped to experiment and invent, then you carry a gift that very few have. Make use of it. Don't worry if your inventions don't work. Your bad inventions may make you disappointed, but they will help others learn and make better inventions in the future. In the end, humanity wins. Progress is made. You have made your contribution.

I hope that I have come across as being humble in all that you have read so far. It's a different matter that the opinion of humble people is not heard, but that's just my humble opinion. Anyway, it is said that science teaches you to be humble. Using that learning, we can say that science has not figured everything out as yet. There's so much that we don't know. The limitations of our scientific achievements, so far, bother us for sure. But they never discourage us. Every day presents an opportunity to try, discover, and learn something new. We make steady and engaging progress along the way. That's what drives us all. A very important fact about all of this is that science is for everyone. It does not create classes, does not discriminate and is not dependent on one person's interpretations and advice. Science is based on measurement and inference, which we can all look at with open access to swim in the great ocean of truth and experience things for ourselves, instead of being made to sit on the beach and told stories about the waves and the waters. That is what religion does. Here we go again!

Cassandra was a figure in Greek mythology. She was the daughter of the king of Troy. For 'rejection of romance' related reasons, she was cursed by God Apollo to utter prophecies that were true but which no one would ever believe. So, she had the knowledge to predict the future correctly but no one gave her predictions any importance. I see science's prediction of the fate of humanity in the same light. Science knows and has told us that we are headed for extinction one day. You should know that a random catastrophic event in nature about 75,000 years ago had very nearly made the whole of the human species extinct. Random events may strike again in the future in the form of stray asteroids whose path coincides with ours. Science has figured out that the Sun will die one day in a few billion years from now, in line with its life cycle and the laws of the universe, and the Earth will be completely scorched to its core in the process. But, humanity is not taking any of this seriously. Our efforts to travel to other planets are minimal, at best. Our focus is completely on unimportant and trivial matters. Our space research efforts and funding are not at all up to the mark. It is funny that we have no regard for countless lives lost in disputes and wars, but suddenly find it absolutely unethical to send men to space without their return being guaranteed. We are born travelers and explorers. We were always meant to visit and live on other planets. It is said that those with wisdom look for a way out. We need to make this a purpose—that we keep our options open.

Elon Musk is a science pioneer who has taken upon him the responsibility of humans colonizing the planet Mars through his venture SpaceX. This multitasking entrepreneur plans to get cargo flights there by 2022, and a crewed flight in 2024. He, without a doubt, is taking humanity forward through the route of science. After having done some amount of reading on him, I have reason to believe that he has unearthed some other little truths about life as well. For example, he is known to skip breakfast on most occasions, in line with all other species on this planet that eat only when hungry and are not made dummies by sellers creating stories about the importance of breakfast but actually wanting to create the market need for cornflakes and oats. Elon spends less time in eating (5-minute lunches) and more time in digesting sitting in the right postures, again in line with other species. Healthy sounding humans like to go on a walk after heavy meals, which is the worst thing one can do after eating. We, conditioned humans, feel that eating slowly helps digest food better, but all it does is make us eat more. Our stomachs, once upon a time, used to digest raw meat, and are more than capable of digesting the juicy and tender things we eat these days. Our stomachs contain acid that is strong enough to dissolve metal, for crying out loud. Mr. Musk also does not take phone calls, and I don't think I need to tell you why that is a great thing that he does. He also likes to break up his schedule into 5-minute slots with tasks for each, thereby allowing him to be in the moment and focused. It is obvious that he has figured out all these little truths in the pursuit of his quest. We should be like him.

I wanted to avoid the repetitive occurrence of the word truth but could not find a synonym for it in the dictionary. Quite rightly so, as the absolute grand truth must be singular and would not come in any other versions. There is a famous Latin phrase, which reads "Vi veri veniversum vivus vici," and translates to “by the power of truth, I, while living, have conquered the universe.” Indeed, every moment is to be experienced with nothing but truth and devoted to the quest for it, through the method of science. If you like watching mystery movies or playing detective games, then you were born to do science. Here, we are solving the ultimate mystery of creation and employing science to find the clues. Think of it like digging a tunnel. Those of us who are better read than others have bigger shovels to dig with. Absolute beginners can use their bare hands to begin with. We’re all in it. The quest is a great leveler and all are equal. We all want to know what’s at the bottom of it and we all dig together. Along the way, we come across things that make us wiser and closer to what lies beneath. Why won’t you join this ultimate team game?

Another good thing about science is that it is a cumulative study, in the sense that no one needs to start from scratch. One can just pick up from the point others have got us to, and build on to strive deeper. So much work has already gone in that it’s now easier to just understand the basics of things and get going. Of course, there will always be detractors who’ll say that science cannot explain the first cause that created the cosmos, so what good is it? Yes, the handicap that science lives with is the lack of evidence available to explain the first cause. Without that, the masses choose to see things like physical laws as divine commands. But, the truth takes time and we should give credit to science for admitting its ignorance instead of proposing unreal answers. The first cause that created creation is one big aspect of the ultimate truth that we are out to find. We’ll find it together. There will be detractors on the way but we can win their hearts and minds. What’s that? You’re thinking that what is the purpose of science if there is no grand purpose of existence? Well, science has helped us discover that truth as well.

The only way to get smarter is to experience more things, to ask more questions and not arrive at conclusions. If our questions are centered on what and when, then we'll only get information. When we start asking questions about how and why, then we start seeking knowledge. Also, we will never have a shortage of various kinds of answers available for questions that cross our minds. We need to choose the right ones. The ideal state, of course, is that when all the 'what', 'when', 'why' and 'how' questions have been answered correctly, and the word 'may' has no meaning. But reaching that state will require a lot of profound discoveries, and the process of discovery is not a predictable or certain process. We need to be patient and at it, all the time. Be open to unlearning. Be vulnerable. Keep our minds ready to be overwhelmed by the things that wait to be unraveled.

The novelist Justin Haythe has written, "A man cannot unsee the truth. He cannot willingly return to darkness, or go blind once he has the gift of sight, any more than he can be unborn." There's no doubt that he has not succumbed to anything untrue. Once you have decided not to conform to conventions, don't give in. You would want to, at many times, but just don't. The rain of curiosity falls all the time, and it is time we got drenched. What an opportunity it is to exist as a human being with a mind that is capable of perceiving and understanding so much. We should make the best use of it. We are the most advanced of all life forms and the pinnacle of evolution's accomplishments. We live in a golden age of discovery and have so much accumulated & documented wisdom from great thinkers and visionaries of the centuries gone by. We also know that the intellectual pleasure derived from demystifying the cosmos is unparalleled. Yet, most of us continue to waste the opportunity. Let's not, anymore. It starts now.

If the beauty and mystery of the cosmos do not excite you as much, and wisdom is something you don't want in great quantities, then that is absolutely all right. No sweat. You can live life doing what you are interested in doing. If painting is your thing, go paint as much as you like. If you like sports, get into that field. Use your life to follow your genetically ingrained and/or upbringing ingrained passions. But pursue only what is true. Don't waste your life believing in society-fed foolish concepts. Unlearn, as required, and stay away from social conditioning. See things for yourself and believe in what is real. Take everything into consideration, be practical and be your own judge. Let me repeat what I had mentioned earlier while defining science i.e. employ common sense, rational thinking, reason-based approach, evidence-based conclusions, well-researched assumptions, prejudice-free inferences, child-like imagination, and wonder. Also, remember that the museum we're visiting has some rules. You cannot do things that cause hurt to others or are against the laws of nature. Be intelligent enough to stay away from those. Stay enlightened that life has no real purpose in the grand scheme, and you are here to make full use of your time here doing quality things that give you intellectual satisfaction, and possibly help you learn more of this incredible reality we find ourselves in.

The universe just is. It is not part of any grand design and nor are we. We are a result of chance and natural evolution. There is no purpose that you and I have in the bigger picture. Let us rejoice this discovery. Nothing can be more liberating than this. Do not fight to deny it. Celebrate it for it frees us and enables us to reach a state of contented acceptance. Welcome every moment of existence. Go do your thing, pursue what you like. Feel no pressure or obligation to live up to any expectations or make things better. There is no deadline to meet. There is nothing to gain and nothing to lose. You don't need to be an ideal son, brother, father, employee or ideal anything. Others will manage if things don't happen perfectly on account of your human shortcomings. It's okay to have imperfections. Make the most of your existence in whatever way that makes it fulfilling. Attain wisdom and make it a point to share what you have. Be good to others and show them the light. And, as recommended, always wonder about your existence and the world around you.

Cheers to the magnificent cosmos we all live in.

Like all good things, this book has come to an end. You are hereby requested to write down your appreciation notes and/or swear words for the author. Use the space provided below. Once done, kindly take a photo of this page and send it to the author on Twitter @sapienRajesh

Bonus material:
Musings on a random day:

Something than nothing perplexity,
A metaphysical profound property,
Of spacetime with finite gravity,
Its dimensions and their invisibility,
Incomprehensible yet a proven entity,
Time and time again, through eternity.
Proof of soul, of maker, of creativity,
A curse, a fallacy, there's no deity,
Morality, belief and revered oddity,
A heresy, of tepid incompressibility,
Of satanic sacrilege and prophecy,
Time after time, through eternity.
A possibility, though infinitesimal probability,
A simulation is what you may see,
Righteous path, sin, such mediocrity,
A heaven awaits, so lame, quite stupidly,
Meaning of life, of forsaken prosperity,
Time after time after time, through eternity.
Truth, mellow and blurred visibility,
A microcosm of prolonged brevity,
The perfection, such fake crudity,
No goodness, no evil, no cruel and no pity,
It's pointless, crystal clear futility,
All through eternity.

Made in the USA
Monee, IL
04 November 2026

67374721R00134